Proof Mind Creates Matter

Why That's Good News for You & Me

by

Stephen Hawley Martin

THE OAKLEA PRESS

WWW.OAKLEAPRESS.COM

www.OakleaPress.com

Table of Contents

Chapter One: The Proof

Proof that mind creates matter came to light more than 25 years ago. I first read about it in *Newsweek* magazine in 1995, but it appears that scientists have either kept this knowledge to themselves, or they have purposefully ignored it, and it's not surprising why. Since at least the middle of the nineteenth century, the majority of scientists have maintained that all that exists is matter, and that consciousness, and therefore mind, is created by it. They have been keeping the truth a secret, which is that the exact opposite is true. The fact is that mind creates matter, as this book will show. To go public with that knowledge would not only be a huge embarrassment to them, it would utterly destroy the credibility of thousands of scientists who have been teaching and writing books that have things backwards. While it's understandable they would not like to have that come to light, it's also a shame because once the truth is widely known, along with the implications of the truth—which are huge—it's highly likely, I believe, that the world will become a better place, and each and every human alive today will have good reasons to adopt an optimistic outlook on life. I will explain why, but first I will lay out the evidence that mind creates physical reality, and that you and I—indeed everything—is consciousness or mind at the core.

Here's the Proof

In 1995, Raymond Chiao, a Hong Cong native and quantum physicist then teaching at the University of California at Berkeley, published a paper about a series of experiments he had conducted. The paper, which, as mentioned above I learned about in the June 19 1995 issue of *Newsweek* magazine, said that what researchers knew or did not know about certain aspects of each experiment had a predictable determination on the outcomes of those experiments. In other words, what was in the researchers' minds apparently determined the results.

The experiments had to do with light. They were variations of an experiment that was first conducted almost 200 years earlier.

Scientists have known for more than a hundred years that light can behave both as waves and as particles of matter (photons), but until 1905 they thought it was composed only of waves. Thomas Young (1773-1829) demonstrated in 1803 that light is waves by placing a screen with two parallel slits between a source of light—sunlight coming through a hole in a screen—and a wall. Each slit could be covered with a piece of cloth. These slits were razor thin, not as wide as the wavelength of the light. When waves of any kind pass through an opening that's not as wide as they are, the waves diffract. This was the case with one slit open. A fuzzy circle of light appeared on the wall.

When both slits were uncovered and the light shown through, alternating bands of light and darkness appeared, the center band being the brightest. Scientists call this a zebra pattern. The areas of light and dark resulted from what is known in wave mechanics as interference. Waves overlap and reinforce each other in some places and in others they cancel each other out. The bands of light on the wall indicated where one wave crest overlapped another crest. The dark areas showed where a crest and a trough met and canceled each other out.

In 1905, Albert Einstein published a paper that revealed light also behaves as though it consists of particles. He did so by using the photoelectric effect. When light hits the surface of a metal, it jars electrons loose from the atoms in the metal and sends them flying off as though struck by tiny billiard balls. So, Thomas Young demonstrated light is waves, and Einstein demonstrated it is particles. What Raymond Chiao's experiment demonstrated is that mind—in this case, a researcher's mind—can change waves of light in to matter—in this case particles known as protons.

Dr. Chiao constructed a double slit experiment in a way that was able to determine what happens when those conducting the experiment observe or do not observe which slits the photons of light pass through. A device was used that fired one photon at a time. Having read about this experiment in *Newsweek,* in July 2008, I reviewed the facts of the experiment as they are presented here with a guest on an Internet podcast I hosted at the time, the noted quantum physicist Henry P. Stapp, author of, *MINDFUL UNIVERSE: Quantum Mechanics and the Participating Observer* (Springer, 2007).

He indicated I understood the facts correctly.

Both slits were open and a detector was used to determine which slit a photon passed through. A record was made of where each one hit. Only one photon at a time was shot, so one would have to suppose that there could be no interference. This was the case. The photons did not make a zebra pattern. Rather, they made marks, tiny dots, on a screen.

When the detector was turned off, however, and it was not known which slit a photon passed through, the zebra pattern appeared. In other words, without the detector—when the researcher had no knowledge—the particles behaved like waves even though they were fired one at a time.

Imagine the stir this caused among those conducting the experiment. In the *Newsweek* article reporting on this, Nobel Prize winning physicist Richard Feynman (1918-1988) was quoted as saying this is the "central mystery" quantum mechanics, that something as intangible as knowledge—in this case, which slit a photon went through—changes something as concrete as a pattern on a screen. To this I would add, "This didn't just change a pattern on a screen, the researcher's mind, i.e., what the researcher knew turned light waves (potential) into matter (photons)."

How could knowledge change the behavior of particles shot from a gun? Conventional science cannot produce an explanation because a tenet of conventional, materialist science is that the brain creates consciousness, awareness, and thought, i.e., mind, and that means consciousness, awareness, and thought are confined within a person's skull. Since it would be ludicrous to suggest that thought enclosed inside a person's head could be capable of having an effect on photons shot from a gun, it ought to be clear to everyone the tenet must be false. As I have written about in other books, The Division of Perceptual Studies [DOPS] at the University of Virginia School of Medicine has concluded after sixty years of research that the brain does not create consciousness. It is a receiver of consciousness that integrates it with the body. Consciousness is not confined within the skull. It originates somewhere else. As another quantum physicist's theory we will explore indicates, physical reality and everything in it is created by consciousness or mind—a mind that we and everything in the reality we share are part of. Yet, as of this writing, many scientists cling to the tenet, saying there must be two different sets of laws of physics: a small (subatomic world) set, and a macro world (the one we live in) set. Somewhere between these two worlds, the laws of physics must change.

This hypothesis does not hold water. First, it doesn't explain why thought contained in someone's head should change things at the subatomic level, or anywhere else. Second, other experiments refute the contention that two different laws of physics exist. One such experiment involves large (carbon 60) molecules called "buckyballs," so big they can be seen and therefore are part of the macro world. Research shows they exhibit the same quantum properties as infinitely small particles. Another is an experiment conducted in 2002 that involved crystals. It produced quantum ridges half an inch high—large enough to be measured with a conventional macro world ruler.

What makes more sense is what William of Ockham (c. 1287–1347) is thought to have been the first to say, "The simplest explanation is the best." The researcher's knowledge, i.e., his or her mind, causes the waves to collapse into particles. To repeat, when the researcher's mind contains the knowledge, the zebra pattern does not occur. Without the information, the zebra pattern appears.

This was verified by setting up the experiment several ways. In the first, the detectors were in front of the two slits. In the second, researchers placed detectors between

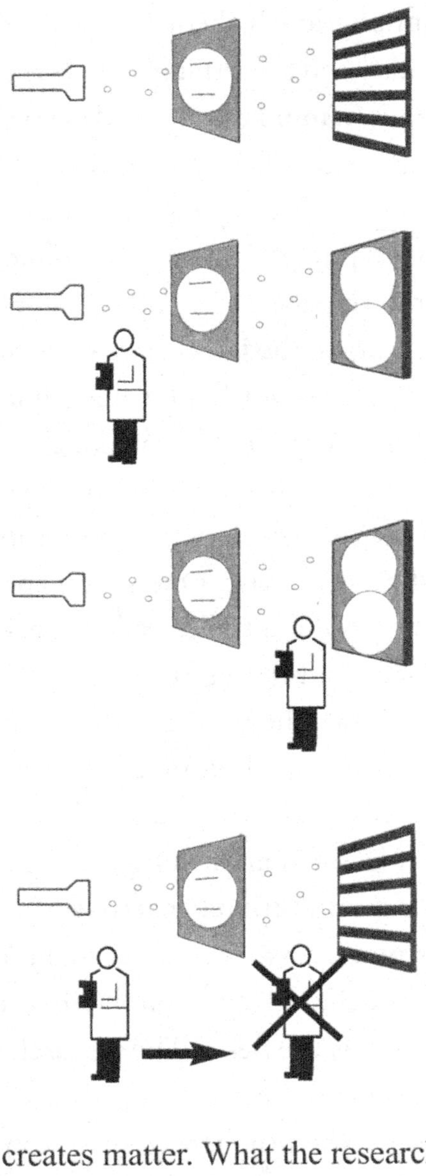

Mind creates matter. What the researcher knows or doesn't know determines whether or not waves of light become photon particles.

the screen and the two slits, i.e., after the photons had passed through the slits. As in the original experiment, knowledge about a photon's behavior at the two slits made the zebra pattern vanish, whether or not the detectors were before or after the slits (see the accompanying graphic). But when the detectors were switched off, and the research had no knowledge, the zebra pattern returned.

In a third variation, a detector was placed before the slits and a mechanism erased the knowledge after the photon had passed through. The same thing happened. The zebra pattern returned. The result was the same no matter which way the experiment was set up—before the slits, after the slits, or before the slits and then erased. Whether or not the researcher knew where each photon hit determined the presence of the zebra pattern, or the lack of it. Versions of the experiment reported on in *Newsweek* were carried out at the University of Munich and at the University of Maryland. The behavior of the photons, the researchers report, "is changed by how we are going to look at them."

The question is, how? As mentioned above, the answer escapes scientists who refuse to think outside the Materialist box, but as soon as one is willing to step outside those four walls, the answer is obvious. Consciousness—mind—creates reality. Before it is observed, reality—in this case a multitude of photons fired through slits—exists as non-material potential in the form of waves. Observed, the waves collapse into matter—photons. What the researcher conducting the experiment believes ought to be the result, is the result—marks on a screen where the photons hit that he or she fired from the photon gun. That fact is the essential truth: mind creates physical reality.

I am not the only person alive today who knows that mind does this. A small number of people do so as well, perhaps most notably, Robert Lanza, M.D., author of *Biocentrism: How Life and Consciousness are the Keys to Understanding the True Nature of the Universe,* and a second book, *Beyond Biocentrism: Rethinking Time, Space, Consciousness, and the Illusion of Death.*

No matter what—whether it's where photons hit a screen or a beautiful sunset—until observed, whatever comes into being exists first as pure potential in the form of waves—a field of them with unlimited potential, what some quantum physicists call "the unified field," and mystics and metaphysicians call "spirit." As the twentieth century psychic Edgar Cayce [1877-1945] said many times while in a self-imposed hypnotic trance, "Spirit is the life, mind is the builder, and the physical is the result." And as was written by Gary Zukav in his book, *The Dancing Wu Li Masters: An Overview of the New Physics* (HarperOne, 2001):

> *. . . the philosophical implication of quantum mechanics is that all of the things in our universe (including us) that appear to exist independently are actually parts of one all-encompassing organic pattern, and that no parts of that pattern are ever really separate from it or from each other.*

To this I will add my belief that underlying this seamless organic pattern—everywhere and in all directions—is mind: What in this book I will call *Infinite Mind.* I know some people of a religious bent are not going to like that term, but my goal is not to convert anyone to a religion of any kind, even though I am a Christian and a follower of Jesus. My goal for this book is to put the truth out there for everyone to see, and here it is: The entire field of all-that-is is waves until a mind, *Infinite Mind,* conceptualizes it, and then the only place we know for certain that it exists at all is within a mind that observes it—your mind and my mind.

More Evidence Mind Can Create Matter

You are about to read about evidence that shows that mind can create matter that was collected by a college professor I spoke with on my podcast named Stephen Braude. At the time, Dr. Braude was a tenured professor of philosophy at the University of Maryland Baltimore County, and I had just read his book, *The Gold Leaf Lady and Other Parapsychological Investigations* (The University of Chicago Press, 2007).

Dr. Braude related several well documented and amazing stories of mind over matter, but perhaps the most fantastic, as well as the one that supports the contention that Infinite Mind creates physical reality, had to do with Katie, a woman born in Tennessee, the tenth of twelve children.

Katie is apparently a simple woman. Illiterate at the time Dr. Braude wrote the book about her, she lived in Florida with her husband and worked as a domestic. She was also a psychic who'd had documented successes helping the police solve crimes. In one instance she was able to describe the details of a case so thoroughly and accurately, the police regarded her as a suspect until those actually responsible were apprehended. She apparently also was able to apport objects—in other words, she somehow caused them to disappear in one place and reappear in another, at least that's what Dr. Braude maintained when I spoke with him. And that was not all. Seeds reportedly germinated rapidly in her cupped hands. Observers claim to have seen her bend metal, and she was both a healer and a medium or channel. Being illiterate, she could not read or write in her native English, but she had been videotaped writing quatrains in medieval French similar both in style and content to the quatrains of Nostradamus. I know Scientific Materialists are going to flip out when they read what comes next because it goes against what they see as a fixed law of physics—that matter cannot be created nor destroyed—but most amazing, perhaps, is what appeared spontaneously on her skin—on her hands, face, arms, legs, and back—apparently out of thin air. It looks like gold leaf, a thin version of the wrapping on a Hersey's Kiss. Katie could not control when this happened, but Dr. Braude and other witnesses saw the foil materialize firsthand. He even videotaped it appearing on her skin.

I just stopped typing and checked. As of this writing, footage from this video can be seen on YouTube. I suggest you have a look. You can use the link below or put "gold leaf lady Braude" in the YouTube search bar. The title of the video is "UMBC In the Loop: Stephen Braude."

https://www.youtube.com/watch?v=iYQ3SpaInk4

Dr. Braude has taken the foil to be analyzed. It turns out not to be gold at all, but brass—approximately 80 percent copper and 20 percent zinc.

Dr. Braude thinks there's a reason she produces brass and not gold. I'll explain shortly, but first let me say that eyewitness reports even by credible witnesses such as Dr. Braude and other researchers who have studied Katie are not going to convince ardent skeptics who cling to materialism. They cannot let go of the materialist tenet that brains create consciousness when in fact brains are simply a way to connect and integrate our consciousness and our bodies so that we can operate in this dimension.

So, where does the brass foil that appears on Katie's skin come from? There is no other plausible explanation: Her mind creates it. In fact, as mentioned, Dr. Braude believes she produces brass rather than gold for a reason. You see, Katie has a difficult and tense relationship with her husband. Once she apported a carving set. It just appeared. And her husband—apparently nonplussed—said, "So what? It's not worth anything." Soon afterward, gold colored foil began appearing on Katie's skin. But it wasn't real gold, it was fool's gold—brass. Dr. Braude thinks this is how she gets back at her husband. Katie's mind—albeit the unconscious part—creates matter in the form of brass foil.

This being the case, why should it be difficult to believe the Infinite Mind underlying everything created us, and the dimension we inhabit? The physical universe had to come from something. One thing quantum physicists agree on is that the physical universe isn't really physical. It is vibrations, and subatomic particles are not particles. As we already know, they are vibrations—waves. Observation turns waves into things, as in the double slit experiment. Mind creates physical reality.

More Evidence

Does mind really create everything? Does it even create biological life? The answer is "Yes," and here's proof. In 1957 Francis Crick [1916-2004] discovered that the chemical subunits along the interior of the double helix of DNA function like alphabetic characters in a written language, or the digital characters such as the zeros and ones in a computer code. No doubt you've seen DNA code printouts. Crick realized they direct the construction of proteins and protein machines that all cells need to stay alive. In other words, it came to light that digital information directs the construction of the crucial components of living cells. Therefore, to explain the origin of life, one would have to explain how this complicated processing system came about.

How complicated is it? According to an article on the website of *BBC Science Focus Magazine,* the UK's leading science and technology monthly: "The DNA in your cells is packaged into 46 chromosomes in the nucleus. As well as being a naturally helical molecule, DNA is supercoiled using enzymes so that it takes up less space. If you stretched the DNA in one cell all the way out, it would be about two meters long and all the DNA in all your cells put together would be about twice the diameter of the Solar System."

How incredible is that? The strand of DNA in a single cell is six feet, six inches long. That's a lot of code when you consider that the size of the characters in the code is microscopic. Think of the enormous amount of information packed into it.

It should go without saying that whenever we see information, and we trace it back to its source, whether it's computer code, a paragraph in a book, or a computer program, there is always an intelligent input, i.e., a mind, that accounts for that information. This indicates, of course, that a mind is behind the origin of life, and yet an ardent Scientific Materialist would argue that given infinite time, anything can occur—for example, that a room full of monkeys with typewriters would produce *War and Peace* or the complete works of Shakespeare with no typos, given enough time.

What argues against this is that the universe began with the Big Bang 13.8 billion years ago and the realization that the earth is only about 4.5 billion years old. Of course, it's true that some scientists argue against the theory that the universe had a beginning. They believe the universe has always existed and that it contracts and expands. But whether it had a beginning or has always existed and contracts and expands, the result would be the same—it got off to a (perhaps new, after an infinite number of previous) start(s) 13.8 billion years ago. This is indicated by a broad range of phenomena, including the abundance of light elements, the cosmic microwave background (CMB), large-scale structure and Hubble's law, i.e., the farther away galaxies are, the faster they are moving away from Earth.

According to calculations made by mathematicians who are a lot smarter when it comes to arithmetic than I am, the odds are much too long for life to have happened by accident, given that the earth is only 4.5 billion years old and life began on earth 3.77 years ago. That leaves 730 million years for the monkeys to have typed out six feet, six inches of microscopic code—with no typos.

I can almost hear someone reading this book thinking: "Yeah, well, maybe life was brought here by a space alien." Maybe. But where did he or she get it? And who or what created the space alien?

It seems to me that anyone who thinks deeply about what is written above would have to come to the conclusion that information resembling computer code that directs something complicated to happen must be the product of some sort of intelligence—in other words, some sort of mind. And yet that would have been impossible if material substance—matter—is all there is because if that were the case, intelligence—a mind—could not have existed until evolution produced a brain.

Not every scientist has turned a blind eye to the facts. After Click's discovery, a number of them began to see that there must be some sort of guiding intelligence responsible for the origin of life. I know this from personal experience because I read a book more than forty years ago that put forth that argument. Published in 1975, it refuted the idea that intelligence, consciousness, and awareness, came about as a result of evolution. The book was entitled *Intelligence Came First.* It was produced by a group of well qualified individuals that met monthly to read and discuss material that was compiled and edited by Ernest Lester Smith [1904-1992], a Fellow of the Royal Society—the prestigious scientific academy of the United Kingdom, dedicated to promoting excellence in science.

Intelligence Came First caused quite a bit of controversy when it came out. Among other things, the book put forth the DNA, computer code argument above, and it also noted that throughout the eons of evolution, needs have preceded the organs through which they are fulfilled—eyes, ears, taste buds, hearts, kidneys, and so forth. Since each new organ developed in response to a need, the book's contributors argued, why would the brain be an exception? The conclusion the book's authors and contributors came to was that intelligence came first, quite able to function in its own realm.

I still have a copy of this book, which has long been forgotten perhaps by everyone except me—because Materialists shouted it down with a vengeance. But think of the intelligence that would be required to design any one of those organs. Could all of them have come about by chance, i.e., random mutations followed by natural selection? Who that's really thought about the complexity of an eye, a liver, or a kidney could possibly think it could have happened by accident? And yet it appears that back in 1975, the scientific community did just that, and apparently many still think that way today.

In 2007, many years after reading that book, I took the opportunity to become the talk show host and producer of an Internet radio show—a podcast—called *The Truth about Life.* For two years I read, and over that period of time interviewed, more than a hundred authors engaged in quests for truth. Among them were medical doctors, parapsychologists, metaphysicians, physicists specializing in quantum mechanics, near death survivors, theologians, psychiatrists, psychologists, and all manner of researchers into the true nature of reality. I don't recall any of these cutting-edge individuals who actually held to a Materialist point of view, except one guest who could not produce any facts to back up his claims. All he could come up with was something to the effect that a particular claim "cannot be so because it goes against what science tells us." I thought, *Yeah, right. "Scientists also tell us matter creates consciousness even though they are unable to determine how, which is why they call it, "the hard problem."*

Darwin's Theory

Darwin's theory is that mutations happen randomly and that those that help an organism survive long enough to reproduce are passed onto the next generation. That makes sense and is likely an important factor in how organisms adapt when their environments change. But how in practice would that lead to an eye, or an ear, or a kidney?

Suppose, for example, that a chance process does result in something that's moving in the right direction for the creation of a kidney—or to back up a bit, to the creation of computer-like code in DNA that makes life possible in the first place? Entropy, i.e., the natural direction of spontaneous change toward disorder, will work against making further progress. In other words, entropy will likely unwind that progress before additional progress can be made.

In summary, the chance-based argument is faulty for two reasons: 1) time works against the chemical synthesis of life, and 2) there is a limit to the amount of time for it to have happened. Using the example of a four-dial bike lock, for example, how likely is it a thief will be able to break the four-digit code? The odds will be against it happening unless he has enough time to sample more than half of the possible combinations. Therefore, when assessing the plausibility of a random search for an informational sequence, it's necessary to assess how many opportunities there are to do so, versus the complexity of the sequence.

With respect to life, it turns out that when you do the math, both in the prebiotic case (the conditions that make life possible) and in the biological case (once life exists, that mutation and selection result in the evolution of life), the complexity of the sequences is so great—making the number of combinations that would have to be searched so large that there is not enough time to have built standard length functional proteins, much less life itself.

In the next chapter a theory of how physical reality in its entirety came about will be explored.

Chapter Two: A New Theory of Everything

Just over 500 years ago, people in the western world thought the earth was flat, that it was located at the center of the universe, that the sun and stars revolved around it, and an anthropomorphic God had created it. Then along came Christopher Columbus [1451-1506], Ferdinand Magellan [1480-1521], Johannes Kepler [1571-1631], and Galileo Galilei [1564-1642], and beliefs changed. The earth was now thought to be round. The sun was thought to be at the center of the universe, and an anthropomorphic God created it.

This remained what people thought until 1859 when Charles Darwin [1809-1882] published *On the Origin of Species.* After the contents of that book were widely known, if you were in academia or were highly educated, the only socially acceptable things to believe were that the universe had always existed and life had come about by accident. Material substance—matter—was all there was, and intelligence and consciousness did not, and could not have existed, until evolution produced a brain. This remained the scientifically accepted worldview well into the twentieth century. The universe was thought to be much smaller than we know it to be today, it was thought to have always existed, and the sun was believed to be at the center of it. Scientists believed that the Milky Way, what we now realize is one of countless galaxies, was all there was to the universe. Moreover, highly educated individuals still believed life had come about by accident when certain chemicals such as amino acids came together in the primordial sea.

Then Edwin Hubble [1889-1953], for whom the Hubble Space Telescope is named, changed how reality is viewed once again by discovering and reporting that countless galaxies exist in addition to Milky Way, that the universe is expanding, and that the sun is definitely not at the center of it. Hubble published his first paper on the relationship between redshift and distance in 1929, again revolutionizing our understanding of the universe and our place in it—although many continue to believe today that intelligence and consciousness did not exist until evolution produced a brain.

Beliefs are, however, in the process of changing once again. A major revolution in our understanding of the universe is occurring, and it is my hope that this book will inspire at least a few intrepid scientists to work out the mathematics, or whatever calculations are deemed necessary to support the new cosmology this book suggests, i.e., a new theory of creation of the universe and life sufficient to convince diehard Scientific Materialists that intelligence and consciousness were in existence long before evolution produced a brain, and that they in fact mind, Infinite Mind, is the source of the universe and life.

Much of the work has already been done. John Samuel Hagelin a Harvard-educated Ph.D. in Quantum Physics has pointed toward such a theory by comparing the unified field, which many scientists speculate is what existed before the Big Bang, to the field of pure consciousness or mind known as "Veda" described in the ancient religious texts of India called the Vedas.

Below is a link to a YouTube video of a lecture by Dr. Hagelin about this:

https://www.youtube.com/watch?v=4u3f7_p1i8c&t=972s

Dr. Hagelin maintains that the unified field and Veda are one and the same. Here are two paragraphs of text quoted verbatim from the write up under the aforementioned video:

> *Two sciences, one ancient and subjective and the other modern and objective, describe manifest creation as an expression of infinite dynamism embedded in the infinite silence of the underlying field. Physics describes this relationship in terms of the unified field and vacuum energy, and Vedic Science in terms of Shiva and Shakti.*
>
> *Similarly, at every level of manifest creation, these two descriptions of nature's functioning correspond exactly. Physics describes three super fields giving rise to the five spin types that characterize elementary particles, the resonant frequencies of the unified field and building blocks of creation; Vedic Science speaks of the three Prakritis or Doshas giving rise to the five Mahabhutas that structure the universe.*

Allow me to cut to the chase and do my best to put into simple, everyday language the new theory that I believe coincides with and is supported by the statements above. I say this because the ancient Rishis of India—sages who lived thousands of years ago—believed that Veda, which boiled down into everyday language is consciousness or mind, is the ground of being of all that is. In other words, the Rishis thought that mind gives rise to the universe.

Based on evidence already presented, it seems to me that Dr. Hagelin that the Rishis are correct. Moreover, the theory supports the Rishis' belief that Veda (consciousness or mind) is the core of each human being and every living thing.

In humans, Veda, or mind, manifests as the Self—the "I AM" or the silent observer at the back of your mind and my mind. But do not misunderstand. The Self (Veda or consciousness) is not an individual's mind. Veda, the ground of being, is what directs your hand to turn a page of this book. It is the decision-maker in each of us that observes the thoughts that arise in the mind and decides which ones to act upon and which ones to dismiss.

It is impossible to understate the magnitude of this realization, the implications of which are huge. A few will be touched upon.

What Existed Before the Big Bang?

Does any of this fit with what the majority of scientists believe today?

Physicists have proposed that the spark of existence had its origin in a quantum fluctuation, triggering an explosive chain reaction—the Big Bang—leading to the still evolving universe we inhabit today. This narrative, however, presupposes the laws of quantum mechanics. As British Biochemist Rupert Sheldrake said in a now banned TED Talk, "[Scientists today say] give us one free miracle and we'll explain the rest.' And the one free miracle is the appearance of all the matter and energy of the universe, and all the laws that govern it, from nothing in a single instant."

Suffice it to say that rather than explaining existence, current scientific theories of the origins of the universe have simply pushed things back to a point that raises the question, "What existed before the beginning?" Could it all have come from nothing? Although that is what many scientists purport to believe, it doesn't make sense. As the song in *The Sound of Music* goes, "Nothing comes from nothing, nothing ever could."

Instead of beginning with nothing, it seems logical that the challenge of explaining existence should focus instead on defining a self-existing ground of being for which no explanation is required. Some physicists have proposed that the true ground floor of reality is the seething quantum realm of particles, forming in and out of existence. While this level of reality no doubt exists, there is no clear reason why the primordial situation should be constrained by quantum physics. A deeper level of explanation seems to be required, and what makes sense to me, as mentioned above, is that consciousness or mind is the ground of being. How seething quantum particles came to be the ground of reality calls out for an explanation, but consciousness can explain itself. A unique feature of consciousness is that it does not appear grounded in anything beyond itself. The conscious self is self-producing in so far that it exists only in and to itself. As René Descartes [1596-1650] famously said, "I think therefore I am." In other words, nothing is required beyond consciousness for existence to be a demonstrated fact.

Diehard Scientific Materialists will have a hard time giving up the idea that the brain is what creates consciousness, but anyone familiar with the research by the University of

Virginia touched upon previously knows that's not the case. It's a shame the findings of that research have not been widely publicized—I suspect because science journalists are afraid of being ridiculed by Materialists—but as previously discussed, the brain is a receiver that integrates consciousness with the body, which might better be called the "mind-body complex" because it is a vehicle or apparatus that allows your consciousness to inhabit and operate in this (physical) dimension. If you don't believe this, go to YouTube and put the following phrase in the search bar: "Dr Bruce Greyson Consciousness Independent of the Brain." A lecture by Dr. Greyson that goes into detail about UVa's research should appear at or near the top. Dr. Greyson, by the way, is not some New Age Looney Tune. He is Professor Emeritus of Psychiatry and Neurobehavioral Sciences at the University of Virginia School of Medicine.

All Is One

If the brain does not create consciousness, where do you suppose it comes from? As stated above, consciousness is the source, Infinite Mind, the Veda or ground of being that generates and informs reality. Let's consider evidence that supports this. The British quantum physicist, Sir James Jeans [1877-1946], wrote the following: "The universe begins to look more like a great thought than a great machine." He was not only onto something, he hit the nail on the head. He knew what all quantum physicists know, that there is no such thing as matter per se—no separate things made of truly solid stuff. All that exists are vibrations—energy, pure and simple. And since physical reality is vibrations, nothing is truly separate—everything is connected.

There can be no barrier or edge where one vibration stops and another one begins. As mystics have been saying since the dawn of time, "All is one."

Moreover, if consciousness or mind is primal, mind must create everything, including the building blocks of space, time, mass, and charge. As Sir James Jeans seems to have suggested above with his comment about the universe resembling a thought, the physical reality we inhabit might be compared to a thought or a dream Infinite Mind is having.

Consider this. When you and I dream at night, our minds create our dreams, their imagery and all their trappings, and yet they seem completely real. If a car in your dream is about to run you over, for example, you are certain it's going to hurt, which is why you wake up. Yet not until your eyes pop open, do you realize it was just a dream.

Now ponder this. There has to be a character in your dream from whose point of view what's happening is observed. There cannot be a dream without someone observing it, and in your dreams that character is you. It follows that if the universe and this world is a "dream" in the Infinite Mind. In your dreams and my dreams there has to be a character in the dream for it to be observed. Who is the character?

You are that character, and I am that character.

We humans—all living things—are characters in Infinite Mind's dream that observe what's going on. And since all is one and connected, including Infinite Mind and us, we serve as the Creator's eyes and ears. We are the vehicles by which Infinite Mind knows his/her/its creation.

The reason this may be difficult to accept is that in our culture we are used to thinking of ourselves as totally separate entities. We think of the universe as "out there." You think of this book, or the Kindle device or phone you are holding, as a separate object. But as noted above, quantum mechanics says that isn't the way things really are. We are all one, all part of the dream. This will begin to sink in and make sense if you do not dismiss it out of hand and truly think about it, objectively.

Panpsychism Is Off the Mark

Let me say here parenthetically that the theory being put forth in this book differs completely from the one called "panpsychism," which some Scientific Materialists are promulgating in an attempt to explain how matter creates consciousness. Panpsychism is the idea that everything material, however small, possesses an element of consciousness—that at the core, even an atom or a quark, photon, or an electron has a tiny level of consciousness.

According to this theory, consciousness is linked to information processing—the more information being processed, the higher the degree of consciousness. This would explain why a human has a higher level of consciousness than a mouse or an ear of corn. Based on this theory, complex information processing activity would in turn create complex consciousness.

However, based on research and observations by DOPS researchers at the University of Virginia School of Medicine, the panpsychism theory cannot be valid. To repeat briefly what the UVa researchers have found, the brain is a receiver of consciousness, it does not create it. How else could children remember past lives, given that between lives they had no brain? How could people be conscious of what doctors and nurses are talking about, and how could they observe what is going on in the operating room, when their brains have flat-lined and they are clinically dead? When the brain has flat-lined, it cannot be processing information. If you want to know more about this, read my book, *Afterlife, The Whole Truth: Life After Death Books I & II,* the bottom line message of which is that there is substantial evidence that, as Pierre Teilhard de Chardin [1881-1955] famously said, "We are not human beings having a spiritual experience. We are spiritual beings having a human experience."

Here's a link:

https://amazon.com/dp/B07J46QQW8

Only Consciousness Is Conscious

It bears repeating that in an attempt to find a solution to the mystery of creation, Materialists have been attacking it from the wrong direction. Physical reality is generated by and contained within mind, not the other way around. Matter does not create consciousness. Consciousness creates matter, and therefore, consciousness is the place to begin an investigation.

To get one's mind around this, it may be helpful to realize that it is impossible for anyone or anything to experience physical reality directly, which is an indication physical reality does not exist and cannot exist outside of consciousness. We experience this dimension through sight, hearing, touch, smell, and taste—the five physical senses of our body-mind complex. We cannot experience any physical thing without using at least one of those senses. Moreover, our consciousness and Infinite Mind—Veda—are one and the same—the "I AM" in each of us.

Chapter Three: How We Got Were We Are Today

One reason it's hard to accept that we are all part of one mind, that consciousness is nonlocal, and that consciousness continues after death, is that it requires scrapping the materialistic model of the world many still hold. But it's important to understand models are just that—approximations of how things are. Whether we realize it or not, we each have a worldview—or a model of how things work. As I have written in other books, you might think of a worldview or model as a stack of cans that forms a pyramid you might see as a grocery store aisle end display. Each can represents an individual belief. Each belief in the display supports other beliefs. Try to change a foundational belief, and the whole thing may come tumbling down. In the past 40 to 60 years scientists have been presented with information that will require many of them to tear down and rebuild from the ground up the model they hold of reality. Up until now, ardent Scientific Materialists have taken the easy way by dismissing as anomalies whatever information does not fit their worldview. Enough of these so-called anomalies have now built up that they might be compared to water that has backed up behind a dam. So much is now there that the time has come for the dam to burst.

If you are one of those scientists, it may be helpful to take a look at how we got to the worldview, or model, that's about to be washed downstream.

A Very Old Worldview

There was a time, anthropologists tell us, when humans felt at one with nature. This can still be seen today in primitive cultures. We humans felt we were an integral part of the ecosystem. The divine showed itself in many forms and was present in all things.

But as humans grew more self aware, they began to feel separation. The myth of Adam and Eve recalls the time when humans parted company with the view that they could commune with the divine. They cut the cord by exercising free will.

No longer seeing the divine in themselves and in others, we humans conjured up gods outside ourselves. In ancient Greece, for example, many gods representing various human qualities were thought to exist. The worldview that evolved in those ancient times had man in the middle between two worlds—a place the Chinese referred to as the Middle Kingdom. The gods lived above the clouds of Mt. Olympus, although they did come to earth now and then, mostly to cause problems for humans.

Below the Middle Kingdom—what caused it to be in the middle—was the underworld, home of the dead, where Hades was in charge and the three-headed dog Cerberus guarded a gate one got to after crossing the River Styx.

Different cultures had different takes on this three-layered universe. Then as now, ideas about the divine and gods differed depending on the group one belonged to. The Egyptians had Bal. The Jews had the god of Abraham. The Romans and the Greeks had a pantheon full.

The Idea of One God Evolved and with It a New Worldview

Then along came Jesus of Nazareth and the idea emerged that only one God ruled over creation—although He did have angels and eventually saints who took up some of the positions left vacant by departing Roman and Greek gods.

In 1994 Karen Armstrong published a book, *A History of God: The 4,000-Year Quest of Judaism, Christianity and Islam* (Ballantine Books, 1994), which chronicled the history of the emergence of the one-God concept. Because of this idea, the worldview changed somewhat. God and angels replaced the pantheon of gods above the clouds. A fallen angel, Satan, replaced Hades. The place below the ground became hell rather than the underworld—where evildoers went. The good folk would be raised at the end of time on judgment day and be given new, light bodies.

This view held sway for better than a thousand years in Europe and Western Culture but was destined to change again because of a new scientific discovery by Christopher Columbus (1451-1506).

Columbus lived on high ground overlooking a Mediterranean harbor. I have visited the ruin of what is said to be the house where he grew up in Corsica, then part of the city-state of Genoa. In that part of the world there is almost no humidity, and the air is very clear. If Columbus had good eyes, he would not have needed a spyglass to see ships climb up over the horizon as they approached the harbor. I've witnessed this myself. Columbus could see the world was round, and he must have decided to prove it by sailing west to get to the spice islands of the East Indies.

Columbus never realized it himself, but he didn't actually get there. Nevertheless, some of Ferdinand Magellan's (1480-1521) crew did, and beyond. Of the 237 men who set out on five ships in 1519, 18 actually completed the circumnavigation of the globe and returned to Spain in 1522.

The newly realized fact that the world was round forced the then commonly held worldview to change. Nevertheless, since people and, most important, Church leaders believed that God had created it, the earth remained at the center of the universe. Now heaven, the dwelling place of God, was seen as being somewhere above the stars. Hell was still beneath the ground, down where it was hot, the place from which molten lava spewed when volcanoes erupted.

The Worldview Gets an Update

It wasn't long before this worldview had to be updated. A fellow named Nicolaus Copernicus (1473–1543) determined the sun was at the center of the solar system. But the Church—the authority back then as science is today—pretty much ignored this concept because it did not go along with accepted canon. This "look the other way" tactic has been used time and again in the twentieth and in the twenty-first century by Scientific Materialists.

A century later, along came Galileo Galilei (1564–1642), a man who would not leave well enough alone. Galileo—among other things an astronomer—championed Copernicus's assertion as proven fact. As a result, Galileo started having to watch his back. This was heresy. At that time people were being burned at the stake for less. Indeed, the leaders of the Church told Galileo he'd better recant, and he did. As a result, Galileo got off easy, spending the final years of his life under house arrest on orders of the Inquisition.

But even the Church couldn't keep word from getting out. Gradually, the accepted views of the day began to change.

A Tiny World Comes into View

In 1675, a Dutchman named Antoni van Leeuwenhoek (1632-1723) —an amateur lens grinder and microscope builder—saw for the first time tiny organisms he called "animalcules" living in stagnant water. He also spotted them in scum collected from his teeth. Leeuwenhoek didn't know or even speculate that "animalcules" might cause disease. It took until the nineteenth century for that revelation to dawn. At the time, the idea that creatures so small they were invisible to the naked eye entered the body to make a person sick and sometimes die would have seemed totally absurd. It was thought demons and the devil caused such things, or that God did it to punish sinners. In 1692 in Salem, Massachusetts, 18 were hanged and one was crushed to death because they were thought to be witches in league with Satan. No wonder after that, and down until today, the idea of Satan and demons and witchcraft was thought to be pure superstition. To believe in such things was to invite witch-hunts and mass hysteria, and nobody wanted that.

The Age of Reason Dawns

Even so, a new day was dawning, a period alternately referred to as "The Age of Enlightenment" and "The Age of Reason." Years before, the English philosopher, Thomas Hobbes (1588-1679), had argued that aside from God—the "first cause" who created the material world—nothing existed that is not of the material world. The logic he used was simple. How could it if God created everything? Of course, at that time, all anyone could see was material stuff.

This view was ultimately to lead to the Clockmaker Theory, the idea that God created the universe, wound it up, let it go, and was no longer involved in its operation. Natural laws also had been created that continued what had been set in motion. Called Deism, many founding fathers, including my personal hero, Thomas Jefferson, subscribed to this view.

Hobbes had a big impact on the Age of Enlightenment, which was to pick up steam in the eighteenth century. But the big kahuna was Sir Isaac Newton (1643-1727), an English physicist, mathematician, astronomer, natural philosopher, alchemist, and theologian. Certainly one of the most influential men of all time, his *Philosophiæ Naturalis Principia Mathematica,* published in 1687, is considered to be the groundwork for most of classical mechanics. Newton described universal gravitation and the three laws of motion that dominated the scientific view of the physical universe at least until the advent of quantum mechanics. It seems safe to say Thomas Hobbes's materialistic view of reality coupled with Newton's mechanistic view is the bedrock of most scientific thinking today, except among quantum physicists.

The prevailing worldview that emerged from the Age of Reason was that the universe might be compared to a giant machine. The Sun was at the center of the solar system. The Earth and planets revolved around it. Nothing existed but the material world. What was thought of in the seventeenth century and earlier as the invisible world of spirit did not exist. Everything that happened had a logical cause. Natural laws governed everything.

Darwin's Theory Takes Hold

In 1859 an Englishman, Charles Darwin (1809-1882), published *On the Origin of Species,* a seminal work in scientific literature and a landmark work in evolutionary biology. Its full title, *On the Origin of Species by Means of Natural Selection, or the Preservation of Favoured Races in the Struggle for Life,* uses the term "races" to mean biological varieties. Darwin's book introduced the theory that populations evolve over the course of generations through a process of natural selection. It presented a body of evidence indicating the diversity of life arose through a branching pattern of evolution and common descent. In other words, God had not created the variety of life on the planet, nor had He created humans. All this had happened through a natural—what might be seen as a mechanical—process. This became accepted as fact among the educated classes.

Darwin's theories reinforced the rationalist idea that the so-called supernatural was a figment of human imagination and—not wanting to be burned at the stake, figuratively or literally—most scientists probably wanted to keep it safely buried. Life and its diversity were results of a natural process known as "Survival of the Fittest" coupled with the environment in which a particular species had evolved. Intelligence and mind had evolved as life had evolved and had reached its pinnacle in humans. Mind and intelligence were produced by an organ, the brain, which had resulted from this evolution. Thought was created by the brain and would later be envisioned as being a result of electrons jumping across synapses. It was contained within the skull. ESP was impossible and so was magic.

A Wedge Between Science and Religion Is Hammered in

With this worldview, a wedge was inserted and hammered in between science, religion and any possibility of things so called supernatural. Hobbes had said nothing existed but the physical. If this were so, where could God possibly reside? What about the heavenly hosts? Thought was contained within the skull so what possible good could prayer do?

A line had been drawn. Educated men and women could not believe in God and prayer or angels or ghosts and demons, which were seen as figments of ignorance and superstition. Many may have had a yearning for God—as humans seem to for the spiritual—but could not rationalize His existence. All were forced to choose between religion and science, though many attempted to straddle the line—as they still do today.

Now, in the first quarter of the twenty-first century, this worldview continues to be the only socially acceptable one in some circles. But there are signs it is beginning to crumble. Hundreds of thousands, perhaps millions, have shifted to a new worldview based on a new branch of science called quantum mechanics and the findings of scientific research that do not fit the materialist-reductionist mold. I hope this book will do its part to knock that view down once and for all.

Pioneers Leading the Way

Let's look at some who have not been afraid to speak out, as well as their ideas and discoveries that conflict with the prevailing nineteenth-twentieth century worldview. The following does not in any way represent an exhaustive list. My apologies to anyone who feels left out, and to anyone who thinks I have overlooked a key figure.

Matter = Energy

In 1905, Albert Einstein (1879-1955), a German-born theoretical physicist, published a paper proving that light behaves both as a wave and as particles. This, as well as Einstein's famous formula, E = MC2, indicates reality and matter are not what they seem. Matter or mass as it is referred to in this formula is equivalent to energy and vice versa.

Quantum physicists came along later and expanded on Einstein's work. Niels Henrik David Bohr (1885-1962), a Danish physicist, made fundamental contributions to understanding atomic structure and quantum mechanics, for which he received the Nobel Prize in Physics in 1922. He is quoted as having said, "Everything we call real is made of things that cannot be regarded as real."

Nothing is really solid. Everything is energy—vibrations.

The Collective Unconscious

In 1912 Swiss psychiatrist Carl Jung (1875-1961) published Wandlungen und Symbole der Libido (known in English as The Psychology of the Unconscious) that postulated a collective unconscious, sometimes known as collective subconscious. According to Jung there is an unconscious mind shared by a society, a people, or all humanity, that is the product of ancestral experience and contains such concepts as the classic archetypes, science, religion, and morality.

ESP and PsychoKinesis Are Proven Real

In the early 1930s a man named J. B. (Joseph Banks) Rhine moved from Harvard University to Duke to set up a parapsychology laboratory. Rhine not only founded the parapsychology lab at Duke, he also founded the *Journal of Parapsychology* and The Foundation for Research on the Nature of Man. His double blind studies conducted largely between 1930 and 1960 established that ESP exists and is real. Not mentioned in our earlier discussion, they also showed psychokinesis—mind over matter—is real as well, at least to a statistically significant degree.

His findings were either scoffed at or ignored by the scientific community then as they continue to be today.

Zen Is Introduced to the West

In 1953, Eugen Herrigel (1884-1955), a German philosopher who taught philosophy at Tohoku Imperial University in Sendai, Japan, from 1924-1929 published the book, *Zen and the Art of Archery* (Vintage Books, 1999). This introduced Zen Buddhism to the West and the concept that "All Is One," i.e., everything is connected rather than made up of separate parts. How else could Zen masters shoot arrows while blindfolded and consistently hit the bull's-eyes of targets many yards away?

In 1966 a British philosopher named Alan Watts (1915-1973) published a book called *The Book: On the Taboo Against Knowing Who You Are* that went into detail about Buddhist thought. Known as an interpreter and popularizer of Asian philosophies for a Western audience, Watts wrote more than 25 books and numerous articles on subjects such as personal identity, the true nature of reality, higher consciousness and the meaning of life. His writings and ideas fueled a new movement which came to be known as "New Age."

Plants Tune into Thoughts

In 1966 a polygraph expert named Cleve Backster (1924-2013) began research that demonstrated living plants tune into the thoughts and intentions of humans as well as other aspects of their environments, thus indicating some sort of hidden mental connection between living things. His findings were ridiculed, but have since been confirmed by other researchers.

Near Death Experiences (NDEs) Are Studied

In 1978 a young man with a B.A., M.A., and Ph.D. from the University of Virginia and an M.D. from Georgia Medical School named Raymond Moody (born 1944) published a book called *Life After Life,* in which he detailed the experiences of people who had been clinically dead and resuscitated.

The Phenomenon of Grace Is Publicized

Also in 1978, a psychiatrist named M. Scott Peck (1936-2005) published a book that became a huge bestseller called, *The Road Less Travelled: A New Psychology Of Love, Traditional Values And Spiritual Growth.* Among other things, Peck's book dealt with the phenomenon of grace. He said grace was both common and to a certain extent, predictable. He also wrote that, "grace will remain unexplainable within the conceptual framework of conventional science and 'natural law' as we understand it."

Grace is the unseen force that brings the best possible results out of unfortunate events and circumstances, .i.e., "every cloud has a silver lining." In Peck's own words, "There is a force, the mechanism of which we do not fully understand, that seems to operate routinely in most people to protect and encourage their physical health even under the most adverse conditions." His book gives specific examples.

Quantum Physics Is Introduced to the Masses

In 1979, Gary Zukav, a former Green Beret during the war in Vietnam, published a book called *The Dancing Wu Li Masters: An Overview of the New Physics*. Targeted for laymen, it explained the basics of quantum physics in everyday language, i.e., without the use of complicated mathematics. Zukav concluded that quantum mechanics indicates, as previously stated, that mystics have been right all along in saying, "All-Is-One."

Also in 1979, James Lovelock published a book called *Gaia: A New Look at Life on Earth* that explained his idea that life on earth functions as a single organism. In contrast to the conventional belief that living matter is passive in the face of threats to its existence, the book explored the hypothesis that the earth's living matter—air, ocean, and land surfaces—forms a complex system that has the capacity to keep the Earth a fit place for life. Since Gaia was first published, many of Jim Lovelock's predictions have come true.

The Spiritual Dimension Is Explored

In the mid 1980s a television series appeared on PBS called *The Power of Myth,* featuring author and Sarah Lawrence College Comparative Religion Professor, Joseph Campbell (1904-1987). These programs made an impact on a significant segment of the public and opened their eyes to the possibility of the existence of what might be termed "a spiritual dimension." This can be summed up using Campbell's own words, "Anyone who has had an experience of mystery knows there is a dimension of the universe that is not that which is available to his senses."

Scientific Studies Demonstrate the Efficacy of Prayer

As previously covered, in July, 1988, Dr. Randolph Byrd, a cardiologist, published an article in the *Southern Medical Journal* about the effects of prayer on cardiac patients. Over a ten-month period, he used a computer to assign 393 patients admitted to the coronary care unit at San Francisco General Hospital either to a group that was prayed for by home prayer groups (192 patients), or to a group that was not prayed for (201). A double blind test, neither the patients, doctors, nor the nurses knew which group a patient was in.

The patients who were remembered in prayer had remarkably, and a statistically significant number of better experiences and outcomes than those who were not prayed for. Also, fewer prayed-for patients died, although the difference between groups in this case was not large enough to be considered statistically significant.

Morphogenetic Fields Are Postulated

In 1994 Rupert Sheldrake, a British biochemist whose theory has already been discussed, published a book called *A New Science of Life.* The editors of the British journal, *Nature,* no doubt being faithful Scientific Materialists, called this book, "the best candidate for burning there has been for many years."

What Researchers Know Can Determine the Outcome

As covered in Chapter One, Raymond Chiao, a Hong Cong native and quantum physicist then teaching at the University of California at Berkeley, published a paper in 1995 about a series of experiments. The paper, reported upon in the June 19, 1995 issue of *Newsweek* magazine, said that what researchers knew or did not know about certain aspects of each experiment had a predictable determination on their outcomes. In other words, what was in the researchers' minds—i.e. thought—apparently determined the result.

Remote Viewing Is Used by the U.S. Government to Spy

In 2001, F. Holmes ("Skip") Atwater published a book detailing how in 1979 he set up and managed—until his retirement from the Army in 1988—a remote viewing unit of U. S. Army intelligence. The book was entitled, *Captain of My Ship, Master of My Soul: Living With Guidance* (Hampton Roads Publishing Company, 2001).

Prayer Adds Fuel to the Life Force

Also in 2001, a study published in the September issue of the *Journal of Reproductive Medicine* showed that prayer was able to double the success rate of in vitro fertilization procedures that lead to pregnancy. The findings revealed that a group of women who had people praying for them had a 50 percent pregnancy rate compared to a 26 percent rate in the group of women who did not have anyone praying for them. In the study—led by Rogerio Lobo, chairman of obstetrics and gynecology at Columbia University's College of Physicians & Surgeons—none of the women undergoing the IVF procedures knew about the prayers on their behalf. Nor did their doctors. In fact, the 199 women were in Cha General Hospital in Seoul, Korea, thousands of miles from those praying for them in the U.S., Canada and Australia. This collaborates with other studies and quantum physics theory that distance is not a factor at the subatomic level of mind.

Research Tells How Best to Pray

An organization exists that has as its purpose the study of what prayer techniques produce the best results. It's called Spindrift and was founded by Christian Science practitioners who have been at this since 1975.

The first question Spindrift researchers sought to answer is, does prayer work? The answer, as we already know, is yes, according to their research. The study give clues on prayer techniques that deliver the best results.

Studies similar to this have been and are being carried out by a consortium of scientists put together by Lynne McTaggart, author of the book published in 2002, *THE FIELD: The Quest for the Secret Force of the Universe, and* her 2008 release, *The INTENTION EXPERIMENT: Using Your Thoughts to Change Your Life and the World.* When she was on my show in early 2008, she described some of these experiments and the terrific success she and her colleagues have had. She said several of these studies were already being prepared for publication.

Mind Is Shown to Create Matter

In 2007, Stephen E. Braude published the book already discussed, *The Gold Leaf Lady and Other Parapsychological Investigations.* The book tells the story of Katie, a woman who demonstrates mind can produce matter—in this case brass: 80% copper and 20% zinc with its huge implications for quantum physics and the origins of the physical universe.

Mediums Can Relate Accurate Information about the Dead

Also in 2008, Julie Beischel, Ph.D. published a paper in *The Journal of Parapsychology* in which she concluded that "certain mediums can report accurate and specific information about the deceased loved ones (termed "discarnates") of living people (termed sitters) even without any prior knowledge about the sitters or the discarnates and in the complete absence of any sensory sitter feedback. Moreover, the information reported by these mediums cannot be explained as a result of fraud or 'cold reading' (a set of techniques in which visual and auditory cues from the sitter are used to fabricate 'accurate' readings) on the part of the mediums or bias on the part of the sitters."

In the upcoming chapter we will discuss what implications of the new worldview for you and me, personally.

Chapter Four: Who You Are

It appears that we have arrived back where we began a long time ago as hunter-gatherers. What they thought of as the divine, and I've been calling Infinite Mind in this book, are the same. There is actually only one life that takes many forms and is present in all things. What this suggests is that you and I are characters in what might be thought of as a dream that Infinite Mind is having—as suggested previously by the British quantum physicist, Sir James Jeans.

I will now explain the true nature of reality as I have come to understand it. I will tell you what I think you should strongly consider doing about it, assuming you want to make the most of life.

The Secret of Life

In other books I have written, I reveal the secret of life and provide what I consider to be indisputable evidence to support it. Simply put, the secret of life is the "urge to become."

Consider life in all its forms. It is constantly becoming. Like all life—the birds and the bees and the flowers and the trees, chipmunks, squirrels, and beavers—we humans are here on Earth to become. As we become, and all life in all other forms becomes, The Source also becomes. That is what the dream of life is about—becoming—in other words, the goal of life is evolution to ever higher states of being, i.e., higher consciousness.

As such, Earth is a school for humans where we come in order to suffer, as Shakespeare's Hamlet said, "the slings and arrows of outrageous fortune," all of which create opportunities for us to learn and grow. This being the case, we have a choice. We can accept what comes our way—whether good or ill—as embodying lessons to be learned, or we can rail against whatever it might be, in which case we will suffer more than would otherwise be necessary. In addition, we will also have to face the same or a similar circumstance in the future—potentially over and over again—until we learn the lesson it was sent to teach us.

Life Is But a Dream

Sir James Jean is not the only one to put his finger on the truth. Arthur Schopenhauer [1788–1860], the Nineteenth Century German philosopher, did so as well. He observed that specific events and the meeting of individuals, which seemed at the time to have come about by chance, in retrospect—particularly from old age—turn out to be essential components in a constant story line. Schopenhauer said that it is as though one dreamer were dreaming a giant dream in which each of the dream's characters has his own individual dream. I believe this is the case, and each dreamer might be thought of as an individual atom of the whole, or alternatively, a spark of the eternal flame—or perhaps a better metaphor is, "a single wave on the ocean of Infinite Mind."

Do not misunderstand. Each one of us is an individual entity. We each have a unique conscious mind, and we each have a unique subconscious mind. But our minds merge with others into a collective mind that we all share. We each have free will and can do things that bring pain and suffering to ourselves and to others. That's where evil comes from. It happens when a being—incarnate or discarnate—believes he or she is separate from the whole and not subject to judgment or karma. As a result truly reprehensible things sometimes come about. But the One Mind that's having the big dream will almost always have subsequent events work out so that at least some good will come from ill-advised deeds. As the Apostle Paul wrote, "In all things God works for the good of those who love Him." (Romans 8:28) Coordination happens effortlessly for the good of all toward the end goal of life in physical form, which is the evolution of humankind and Infinite Mind to ever-higher levels of consciousness and wisdom.

Evolution Is the Goal

As minds evolve, they become more self-aware. Flowers and earthworms possess only subjective or subconscious minds, their own small portions of underlying organizing intelligence. Their "minds" are subjective because they cannot think about themselves. They can only react in a programmed way to the input or stimuli they receive. A dog and to a much greater extent, a human, have both a subjective mind and an objective mind. Their subjective minds keep them breathing and their bodies functioning while their objective minds think about and analyze situations. Unlike the subjective mind, an objective mind can worry and be afraid. This is both a blessing and a curse. It is a blessing that we can plan ahead in order to avoid trouble and thereby eliminate the uncomfortable sensation of worry. It is a curse because fear is a kind of belief—a belief charged with emotion. Since it does not analyze or judge, the subjective mind works hard to bring about what the objective mind believes. A fear is almost certain, therefore, to manifest if it is allowed to continue unchecked—whether or not that particular fear was originally grounded in reality.

As you now know, there is only one mind because at some point, as we have seen, all minds blend into the Infinite Mind. This is why remote viewing is possible. Psychics are able to push their minds past barriers that separate them from the whole. Like the vast majority of water on earth that is connected but divided into oceans, seas, and rivers, we can think of the one mind as being connected but divided into various levels: the universal subjective mind, the collective subconscious mind of humanity, individual subconscious minds or souls, and last but not least the part of each person's mind of which he or she is aware, the conscious mind. Our personal conscious minds in turn are divided into a conscious portion and an unconscious portion that contains the memories of this life and unconscious and habitual programming.

At the core, however, we are all aspects of the Infinite Mind, consciousness itself. We are all linked together, all one at the core. What this means is that you are not the color of your skin, you are not the school or university you went to, you are not your job or your nationality, and most certainly, you are not your hands or your feet, and you are not your body.

"Wait. I'm not my body?"

That statement may raise doubt in you, but you are not something you can observe, and you can observe your body. In fact, the two words, "your body," suggest your body is not you, but rather, something that belongs to you. You can control the actions of your body. For example, you might decide to force it to dig a ditch, and if you do, after a few hours, you might say, "My body aches." And so it's not you that aches—it's your body—your back and your arms.

How about your mind? Do you think you are your mind?

You are not your mind, your mind is something that was built up over time, and, as you will see, realizing you are not your mind will empower you. Consider this. You can observe your thoughts as they come and they go. You may even observe them to a fault when they keep you awake at night. But you, the real you, do not have to be at the mercy of your thoughts, and therefore, your mind. Like Scarlett O'Hara in *Gone with the Wind,* you can decide to think about tomorrow whatever problem has arisen in your mind today.

It's a fact. Anything you can—sometimes figuratively speaking—stand back and observe cannot possibly be you. Moreover, whether you realized it or not before now—whether you have ever exercised this prerogative or not—you can decide which thoughts to pay attention to and which ones to discard. You can decide what to think about, or not to think about. In other words, you can control the actions of your mind, just as you can control the actions of your body, and this means your mind cannot possibly be you.

You may be surprised to learn that you are not your body or your mind, but think about this: There was a time when people thought the world was flat and that the earth was at the center of the universe—that the sun and the stars revolved around the earth. Now, just about everyone knows the earth is round, that the sun is at the center of the solar system, and that our solar system is one of trillions upon trillions.

Once you get in touch with and live your life from the viewpoint of your true Self, the eternal you at the core—the infinite, eternal you—once you truly accept and embrace the secret of life, you will live in peace and be free from suffering and desire. This will not require learning reams of information about spiritual matters and metaphysics. It will simply require setting aside the ego and getting in touch with the still small voice inside. This is what Jesus knew and what he tried to communicate to anyone who would listen. To illustrate, let me tell you the true story of someone I know that learned this secret. I'll call him George.

Leading up to that day, George had been extremely unlucky in love. What he wanted more than anything was to be in love and to share a truly loving relationship, but at the age of 26, he had already suffered one failed marriage and two unfortunate, long-term relationships. The third breakup sent George into a deep and dark depression. He became suicidal, and except for work, he spent two solid months alone in his apartment, much of it in a fetal position. He didn't see anybody. He didn't talk to anybody. The only thing he did to try to bring himself out of his depression was to listen to Eckhart Tolle podcasts and videos.

If you are not aware of Eckhart Tolle, he is a spiritual teacher and best-selling author, a German-born resident of

Canada best known as the author of *The Power of Now* and *A New Earth: Awakening to Your Life's Purpose.* Tolle's essential message is to stay in the "Now" as much as possible. Rather than dwell on past problems or worry about what you might face in the future, the way to achieve peace is to step back from those thoughts and observe your own mind objectively, distinguish between the voice of your ego and your actual situation, and live as much as possible in the present moment—what Tolle referred to as the "Now."

During lunch and other breaks at work, George would find a quiet, secluded place on a bench outside his office building in the campus style business park where he worked. He would eat a sandwich, play Eckhart Tolle videos, and stare up at the sky. After a month or so of watching white, puffy clouds drift by, and listening to Tolle talk, he began to break free from his depression.

One day he was listening to a video and Tolle was saying things a voice in someone's mind might say that would bring them down, such as, "I'm no good. People laugh at me behind my back." The sarcastic way Tolle was saying these things caused his audience to laugh.

"Life is my enemy. It's treating me so unfairly. If only life would just be nicer to me." Everyone present with Tolle laughed again. Eckhart Tolle laughed, too, and this time George chuckled.

Then Tolle said, "If only I could have that one thing—then I would be truly happy!"

Everyone laughed, and it suddenly struck George that Tolle was talking about him. The inner voice Tolle was mimicking was the voice of his ego—the voice he had been hearing in his head saying if he could only find true love, he would be happy. Unless he did, he would simply have to remain miserable and depressed!

This struck George as so obvious it was hilarious, and he began laughing a sidesplitting, keeling-over laugh. For the first time, he saw through the petty game the ego plays and realized the thought voices weren't actually coming from him. The voice in his head was just some sort of automatic, conditioned-into-him response, a learned-mechanism from his past—his upbringing, his parents, his previous relationships. It was an old tape that played over and over, and it had been torturing him for months—perhaps even years—and the amazing thing was, he didn't have to listen to it. He could ignore it, or better yet, he could simply shoo it away.

This insight came to him in an instant. He realized he had fallen victim to a trap, and with that realization, he was free of it. Free! He laughed and the laughter brought an incredible sense of freedom because he was laughing at his ego, and his ego was an automaton that wasn't him. It was a construction built up over years—fueled by memories he did not have to identify with, and didn't have to let bother him if he didn't want them to.

After a few moments, his laughing turned into crying—not sad tears, but tears of joy, tears of relief. The small-self noises of the ego self had been muted. He had simply stepped out of the poor-sad-guy-who-just-can't-find-love, boohoo-poor-little-me role he had unwittingly taken on, a role he despised, and he could now move on. He'd been caught in a trap—a mind mechanism trap that now had collapsed, leaving him to revel in beautiful, peaceful silence.

The Truth, Higher Self

What George finally got in touch with was his true Self, his "I AM" that is pure and simple consciousness—the Silent Observer that was always there at the back of his mind, and is at the back of your mind—at the back of everyone's mind. It is the Creator, the core of each of us that is never stained by experience; never damaged. It does not age. It is not born. It does not die. It always is—the ground of being of all that is.

George realized that to find it, you have to do what Jesus said to do in Luke 9:23-24 (NIV):

> *Whoever wants to be my disciple must deny themselves and take up their cross daily and follow me. For whoever wants to save their life will lose it, but whoever loses their life for me will save it.*

George realized Jesus identified with the "I AM" ground of being in him and that the lived and operated from that place. And so it struck George that to truly follow Jesus, someone had to first die to him or herself, meaning die to the small, prideful ego self because doing so is the only way to identify with and become full of the One Life that so filled Jesus. In other words, one must empty oneself of the small self in order to fill oneself with the Big Self—the Divine Self.

George also realized that personal identity is like a filter we see reality through. It creates the illusion that we are defined by our circumstances and environment, which is why everyone sees reality differently. Everyone views the world through a different filter—a unique lens. The only way to see true reality is to see it from the Big Self, and so the more identified and attached we are to our little identities, the less clearly we see and the more clouded everything becomes. Unfortunately, our judgments, our awareness, our clarity can become extremely clouded, and when that happens, we suffer, and often we cause the world to suffer along with us.

The End of Suffering

Here is the intended message of what is written above: If you want freedom, if you want liberation from psychological suffering, if you want surcease of sorrow and desire, if you truly want the truth, then you must shoo away the small, ego you, which is a false you, and make it disappear. When you do, the Real You—the "I AM" that is your window to the One Mind—will appear.

This is often accomplished through meditation. When meditating, ego thoughts are pushed away, ignored, or simply observed as irrelevant or perhaps even amusing BS by the Real You, the Silent Observer.

If you do not engage in meditation now, I suggest you give it a try. Go on YouTube and watch a few videos that give you pointers. Spend ten minutes to half an hour in meditation every morning or evening—perhaps even both. If you do that for a while, it may become a habit you look forward to. Be aware, however, that at first your ego mind will object and attempt to torpedo your efforts because it fears annihilation. Remember, the ego is in the business of survival, and so it will keep telling you that you are wasting your time. Don't let it have the upper hand. Keep pushing ego thoughts away or ignoring them. Eventually, if you want liberation and enlightenment enough, the Real You will overcome and the ego's screaming out in the throws of death will fade away.

And something else: Depending on your upbringing, it may also be necessary to jettison old baggage stored in the attic of your unconscious mind, which may be holding you back and causing you to react and behave in unproductive ways. Techniques intended to assist you in doing so will be provided in upcoming chapters.

Chapter Five: Awakening to the Real You

That we are all one and joined to one mind at the core of each of us is not a new idea. Alan Watts [1915-1973], a twentieth century philosopher and interpreter of Zen Buddhism, answered children's questions concerning why they were here, where the universe came from, where people go when they die and so forth with a parable about God playing hide and seek. Watts told them God enjoys the game, but has no one outside himself to play with since he is All-That-Is. God overcomes the problem of not having any playmates by pretending he is not himself. Instead he pretends that he is me and you and all the other people and the animals and rocks and stars and planets and plants and in doing so has wonderful and wondrous adventures. These adventures are like dreams because when he awakes, they disappear. Here is some of what Watts wrote:

> *Now when God plays hide and pretends that he is you and I, he does it so well that it takes him a long time to remember where and how he hid himself. But that's the whole fun of it—just what he wanted to do. He doesn't want to find himself too quickly, for that would spoil the game. That is why it is so difficult for you and me to find out that we are God in disguise, pretending not to be himself. But when the game has gone on long enough, all of us will wake up, stop pretending, and remember that we are all*

one single Self—the God who is all that there is and who lives forever and ever.

It will no doubt be shocking to some to think of themselves as God, but I don't think Watts meant that in a literal sense. He was talking about the core essence that is beyond the ego and deeper within than the personal unconscious, the collective unconscious, the archetypes and so on—the part of you that is eternal. As Joseph Campbell [1904-1987] said in the PBS TV series, *The Power of Myth,* "You see, there are two ways of thinking 'I am God.' If you think, I, here in my physical presence and in my temporal character, am God,' then you are mad and have short-circuited the experience. You are God, not in your ego, but in your deepest being, where you are at one with the non dual transcendent."

Your consciousness seems to belong to the person you consider yourself to be because you have memories stored in your unconscious mind. You also have a name and perhaps a job and a history that was created while inhabiting your current physical body. You may think of yourself as American, Canadian, British, or Australian. Perhaps you grew up poor, or maybe your father was the CEO of a Fortune 500 company. The skin that covers the body you now inhabit might be black, white, brown, or some shade in between. Perhaps you subscribe to Christianity, Judaism, or Islam. You might have been a good student, or not so good, a star quarterback on the high school team, or a 98 pound weakling. The combination of all that gives you a sense of identity©it's who you think you are©your ego. But none of that is who you actually are, and your ego is not you.

As George realized, the ego is a construction, and it is not in the business of eliminating your suffering©quite the contrary. The ego is concerned only with its own survival. It wants to be in charge, and as such, causes you to react the way you do to outside stimuli©it is what makes you feel bad, sad, angry, or fearful. Give it some thought and you will realize your ego is not your friend.

You might think of the ego as a computer program like the one in the movie, *The Matrix,* that prevents you from seeing reality as it truly is. It won't allow you to see your true potential, which is virtually unlimited. The ego causes your psychological suffering. But now you know the truth: you don't have to obey the ego's demands to react to situations as you do today. Besides taking advantage of the moment between stimulus and response to stop and think before you react, you can wake up and realize who you really are©you are Infinite Mind in disguise experiencing your creation. You can become whatever you can truly come to believe yourself to be because of what some now call the Law of Attraction, Christians call the Law of Reaping and Sowing, and in the East is known as the Law of Karma. There's a wonderful little book written by a man named James Allen [1864-1912] that drives home this point. It's called, *As a Man Thinketh,* and I highly recommend it.

The Real You

You are an eternal spiritual being having a temporary physical experience, and the best is yet to come. Knowing that ought to bring you a sense of pure joy. Think about that. Savor it.

Knowing you are at one with the All, and that you are the All is a magnificent feeling. Alan Watts, who apparently had achieved this state, wrote that a gut-level, the realization of oneness brings moments of joy that are incredibly intense, that sorrows are looked upon philosophically, and that the sense of union with the universe empowers you. Once you come into harmony with all that is you will arrive in a position to live the totally fulfilling life you were born to live, and you will be positioned to achieve complete self-actualization. Here is a direct quote of some of what Alan Watts wrote about this:

> *In immediate contrast to the old feeling, there is indeed a certain passivity to the sensation, as if you were a leaf blown along by the wind, until you realize that you are both the leaf and the wind. The world outside your skin is just as much you as the world inside: they move together inseparably, and at first you feel a little out of control because the world outside is so much vaster than the world inside. Yet you soon discover that you are able to go ahead with ordinary activities—to work and make decisions as ever, though somehow this is less of a drag. Your body is no longer a corpse which the ego has to animate and lug around. There is a feeling of the ground holding you up, and of hills lifting you when you climb them. Air breathes itself in and out of your lungs, and instead of looking and listening, light and sound come to you on their own. Eyes see and ears hear as wind blows and water flows. All space becomes your mind. Time carries you along like a river, but never flows out of the present: the more it goes the more it stays, and you no longer have to fight or kill it.*

When you know that we are all one and you are it, you will naturally want to "do unto others as you would have others do unto you." Later, I will share some thoughts about how you might go about identifying your unique gifts and talents and how you might determine the most productive and satisfying way to put them to work in service to others. In the meantime, let's consider how your personal reality is created. If your circumstances and reality at present aren't what you would like them to be, let's also look at how you can create the circumstances and reality you would like to have.

It's a simple fact that beliefs—your beliefs and the beliefs of others—create your reality. This is so because belief—true, unadulterated belief—is powerful. The effectiveness of placebos, for example, has been demonstrated time and again in double blind, scientific tests. The placebo effect—the phenomenon of patients getting well or feeling better after taking dud pills—is seen throughout the field of medicine, and belief by a patient that he or she has taken real medicine is what causes it. One report says that after thousands of studies, hundreds of millions of prescriptions and tens of billions of dollars in sales, sugar pills are as effective at treating depression as antidepressants such as Prozac, Paxil and Zoloft. What's more, placebos cause profound changes in the same areas of the brain affected by these medicines, according to this research. For anyone who may have been in doubt, this proves beyond a doubt that thoughts and beliefs can and do produce physical changes in our bodies.

In addition, the same research reports that placebos often outperform the medicines they're up against. For example, in a trial conducted in April 2002 comparing the herbal remedy St. John's wort to Zoloft, St. John's wort cured 24 percent of the depressed people who received it. Zoloft cured 25 percent, but the placebo cured 32 percent.

Taking what one believes to be real medicine sets up the expectation of results, and what a person expects to happen usually does happen. It has been confirmed, for example, that in cultures where belief exists in voodoo or magic, people will actually die after being cursed by a shaman. Such a curse has no power on an outsider who doesn't believe. The expectation and belief causes the result.

Let me relate a real-life example of spontaneous healing that I believe came about because of her belief and that of others. It involved a woman I'd known for quite some time I will call Nancy, which is not her real name.

Nancy is a minister's wife. She's a devout Christian—as firm a believer in her religion as a bushman who'd drop dead from a witch doctor's curse is in his. Some years ago, a lump more than half an inch in diameter was discovered in one of her breasts. Her doctor scheduled a biopsy.

A prayer group gathered at her home the night before this procedure was to take place. Her friends prayed not that the lump would be benign, but rather, that it would disappear entirely.

Nancy is a member of a denomination that takes the Bible literally. In Matthew 18:19-20, Jesus is reported to have said, "Again, I tell you that if two of you on earth agree about anything you ask for, it will be done for you by my Father in heaven. For where two or three come together in my name, there am I with them."

As you can imagine, it was more than two or three. It was a living room full. The next morning, upon self-examination, the lump in her breast appeared to have vanished. But nonetheless Nancy kept her appointment at the hospital where her doctor conducted a thorough examination.

The lump indeed was gone. Not a trace could be found, and the bewildered doctor sent her home.

How could a solid lump of tissue disappear? It melted away due to the potent combination of belief and expectation. We indeed create our own reality.

Jesus also said, "Therefore I tell you, whatever you ask for in prayer, believe that you have received it, and it will be yours." (Mark 11:24) Notice the tense change in this verse. Jesus is saying to believe that you already have what you ask for and it will be given to you in the future. Jesus apparently knew that thoughts are things and that what we believe already exists in the nonphysical realm of spirit as a thought form. Thoughts are things, as we will see, that are ready to materialize on the physical plane.

How are beliefs able to do this? It has to do with the different levels of mind. You might call them lower and higher, or subjective and objective. What differentiates the higher from the lower is the recognition of self. Microbes, plants, worms, and fish possess the lower kind only. They are unaware of self. Even higher animals such as squirrels and other animals of the forest are likely totally unaware of self. This is indicated by the fact that an antelope, for example, does not seem to become angry with a lion when the lion kills and eats one of it's young. Once the lion is out of sight, the antelope simply resumes going about its business of grazing.

Perhaps some animals, dogs and other pets and perhaps dolphins, elephants and whales, have some level of self-awareness. I once had a dachshund that would let me know his displeasure by pooping on the rug when I left him alone for what he apparently considered too long a time. Certainly all humans, even small children, are self aware, and so it appears that the higher variety of self-aware thought is possessed in progressively larger amounts as if ascending a scale. At the present stage of evolution on Earth, humans possess the top level of consciousness, and within each of us are all the other levels down to the subjective, non-dual ground-of-being mind.

Here are the levels of mind as related to me by a professor at the College of Metaphysics in Windyville, Missouri:

1. An Individual's Conscious Mind
2. An Individual's Unconscious Mind, accumulated during the current life
3. An Individual's Subconscious Mind, accumulated during all his or her incarnations
4. The Collective Subconscious Mind of humankind, containing the archetypes and what is sometimes called the Akashic Records
5. The Subjective, Non-Dual Ground-of-Being Mind of the Creator

Level Five, the ground of being, subjective mind, is the organizing intelligence or mind present everywhere that, among other things, supports and controls the mechanics of life in every species and in every individual. It causes plants to grow toward the sun and to push roots into the soil. It causes hearts to beat and lungs to take in air. It controls all of the so-called involuntary functions of the body. And the fact is, it controls a lot more, including all physical and metaphysical laws.

Level Two, an individual's unconscious mind, contains the beliefs that have been established in this life, and like all levels after the conscious mind, the unconscious mind is subjective, meaning it cannot think out side of itself. This is why your beliefs create your reality. Your unconscious subjective mind determines your circumstances and your reality because it blends into and is part of the mind we all share and because of this influences events either favorably or unfavorably based on your beliefs. In addition, your beliefs —whether conscious or unconscious —influence the decisions and the choices you make.

Years ago, I read a series of lectures given in the early twentieth century in Scotland by a man named Thomas Troward (1847-1916) that made a lot of sense. He said the conscious mind has power over the unconscious subjective mind, and the subjective mind creates your reality. I discovered the truth of this firsthand in college when I learned to hypnotize others. I would put a willing classmate into a trance and tell him he was a chicken or a dog. Much to the amusement of my audience, he would then act accordingly.

Hypnotism works because the hypnotist bypasses his subject's conscious mind and speaks directly to the subject's subjective mind. Because of this, a subject's conscious, objective mind is unable to question or disregard the hypnotist's directive. Of course, once the subject emerges from the hypnotic trance, his or her objective mind will take over and will be able to nullify the hypnotist's directive. Nevertheless, while the subject remains in trance, the subject's subjective mind has no choice but to bring into reality what the hypnotist instructs it to do.

The Role of Feelings in Changing Beliefs

If you want to change a belief buried in your unconscious mind, it's important to realize that how you feel about the belief, or you sense of "knowing at a gut level" whether or not it is true, is as important as facts and logic are when it comes to convincing the subjective mind to discard it. So, if you have been brought up to believe and feel, for example, that you are a victim and will never amount to anything, or that people in your family are destined to be overweight, you actually will be a victim and never amount to anything, and you will also be overweight—until, that is, those beliefs change and your subjective mind is reprogrammed.

As stated, the subjective mind cannot step outside of itself and take an objective look. As such, it is capable only of deductive reasoning, which is the kind that progresses from a cause (what is programmed into it) forward to its ultimate end. Having the mind of a deer, a rabbit or a squirrel, it does not stop to question or analyze. This is the same reasoning a criminal might use in committing a crime. He may walk into a room, see a man counting his money, and think: "I need money, so I will take his. Since the man is protecting the money, I will get rid of him. I'll shoot him. He'll drop to the floor. I will then take the money and run. I'll leave by the window." The subjective mind is non-dual.

Right and wrong, good and bad, are never considered—only how to get to the end result.

On the other hand, the conscious mind, being objective and self-aware, can step outside. It can reason both deductively and inductively. To reason inductively is to move backward from result to cause. A police detective, for example, would arrive at the crime scene and begin reasoning backward in an attempt to tell how the crime was committed, and who might have done it.

So, if you have a victim mentality, your subjective mind will filter out all sorts of opportunities that come your way because it determines what you notice and are attracted to out of the literally millions of things you are exposed to each day, and it is determined to make your beliefs come true. Therefore, if your subjective mind is convinced you are a victim, and nothing you do can change that, it will dismiss out of hand all sorts of opportunities that might lead to a better life if you would only notice and take advantage of them.

It follows that if you want to change your life, you must change your beliefs, and this may not be easy. Repetition of the new belief you want to adopt will help, but it may not be enough by itself because, as mentioned above, you have to feel the belief you want to adopt is true.

Toward Higher States of Consciousness

There have always been a few people who believe we are sparks of the Divine that have evolved on Earth because of an innate desire to become perfectly balanced in terms of love and wisdom until we finally merge back into the Source at the end of our journey in the physical plane of existence. If this is true, it means that to move ahead we must find the "distortions" within us—what I think of has harmful beliefs or negative thought structures. Some would say they are "shadows." Whatever term you prefer, you must purge them from your unconscious mind in order to advance.

Since the unconscious, subjective mind does not know what are good beliefs and what are harmful beliefs, and you may not even be aware they are buried there, the first step is to identify the beliefs you need to change. Beliefs are points of view you have about yourself and the world, and points of view can be changed. They are ideas we think about often, and once we have thought frequently enough about one, the unconscious mind assumes it must be important to your survival as the person you think yourself and wish to be. So, the mind condenses it into a belief, once an idea has become a belief, it exists outside of your conscious awareness.

Let's say because of how you were raised you have a belief you are unworthy of love. If someone asked you if you think you are unworthy of love, you likely would say "No, of course not." You might even be insulted if the individual had the nerve to ask you that.

Unfortunately, you cannot simply look into your conscious mind to discover your beliefs. The way to go about it is to observe your own behavior and life situation because how you behave and where you are today are the results of your unconscious beliefs and opinions. As already discussed, in addition to influencing how you react and your choices, your unconscious mind blends with the universal subconscious mind, causing it to bring into your life what you unconsciously believe about yourself and the world outside you.

Let me interject here that a frequent misconception is that your thoughts create your reality. This is not so because thoughts occur only in the conscious mind, and the unconscious mind is the instrument that creates your reality. Only when a thought is given enough significance to create a belief does the thought gain power. Therefore, a thought you consider to be insignificant or untrue is powerless. The truth is, you give all thoughts their meaning, or the lack thereof. As has been said, a subjective mind does not judge whether something is right or wrong, good or evil. You were a blank canvas when you arrived on this planet on which you could paint the life you want, so why not take advantage of that and do so now?

To begin, it is important to understand that your ego is not you and does not have your best interests at heart. The ego's primary goal is its own survival. It is fully aware of your personal reality at all times and constantly judges whether your reality is in sync with your beliefs. It draws upon them to create thoughts and impulses that attract your attention, thereby seeking to reinforce and uphold those beliefs. Therefore, unless you discover what your unconscious beliefs are, the ego will always be in charge, directing your life.

You can determine the beliefs that govern your life by noticing what repeatedly happens in your life. For example, are you a man or a woman who always seems to attract a member of the opposite sex who ends up abusing you? If so, you must have an unconscious belief you are unworthy and deserve abuse. What are other things that consistently go wrong? A buried belief is the reason.

Something else you can do is take note of anything that sets you off or "triggers" you because that is a result of one of your core beliefs. Here's a simple example: suppose your father was critical of you and frequently criticized you and the result as you grew up was that you felt that nothing you did was ever good enough. Now let's say in your current job you have a male boss older than you, and any time he gives you criticism—even though it may be constructive and meant to be helpful—you immediately feel a sense of fear and anxiety. You have just experienced a trigger because you were conditioned by your father from childhood to believe you are incapable of success. The same process is playing out in obvious and subtle ways in every aspect of your life.

If you want to improve your life and move forward in your quest to live every day in the kingdom of heaven on Earth, you need to notice any time you react to a situation. Then think about your reaction and drill down until you identify the core negative belief that brought it about. Once you have put your finger on it, you can begin the process of changing it to what you want to believe about yourself. Once the belief has been expunged, or replaced with one that's positive, your life will change for the better. Some say the energy centers of our body develop blockages based upon such thoughts and perceptions. If one is lacking in some form of self-love or wisdom, that distorted energy will manifest in one or more of the energy centers. In order to unblock these energy centers to become balanced, we must heal the distortions within us.

I'm aware of two ways this can be done, which are known as the "feminine" approach and the "masculine" approach, the feminine being the positive and masculine the negative polarity — the positive being love and the negative being wisdom. This does not mean that love is good and wisdom is evil. Both polarities are equally valid expressions of the Creator, and that achieving the balance between love and wisdom within the self is crucial for spiritual evolution. Both approaches, while different, can be effective. The feminine heals through transmutation, and the masculine heals through recognition. Which one you favor is up to you.

The Feminine Approach to Balancing

The feminine approach to balancing is accomplished through feeling, thereby purging the unwanted thought form through love. It involves going to the root of the negative emotion, feeling it completely, and allowing it to express itself. Rather than meeting it with resistance, the key is to meet it with love and acceptance, and then, although it may be painful, review and to an extent relive the experience that caused it.

The feminine approach is much more painful and intense than the masculine approach, but it is also much more powerful and immediate. A single healing session has the potential to purify an old wound that has been festering for decades. It takes time, practice, and courage to develop the skill of locating repressed emotions and the memories that are their causes so that you can face them and feel them, but the power of the distorted beliefs and points of view they brought about will be healed quickly as a result.

The Masculine Approach to Balancing

Attaining wisdom and understanding is the masculine approach to healing. Illusions are banished by perceiving that that is simply what they are—illusions. Unlike the feminine, which seeks to heal repressed energy by feeling it out of existence, the masculine approach seeks to defuse and disperse the energy out of existence. It is not as immediate and powerful as the feminine approach, but it is less painful and intense. It is the one I personally prefer, and the one that works best for me. What is required is a permanent shift in perception or attitude, which over time will drain a debilitating thought form of its power.

Here is how I suggest you begin. Step back and observe your own behavior. Once you become aware of what is going on, and you are conscious of how you are reacting in different situations, you will have the power to change. To employ this technique successfully, my advice is to put all your energy and passion into your desire for change. You must want the new belief to be true with your entire being. A fraction of a second exists between the moment something happens and your "triggered" reaction. Stop yourself at that instant. Then change what you think and how you react to conform to your new beliefs—even if at first this feels a little awkward. This will take energy away from the old beliefs and direct it onto the new ones, and it won't be long before you start to feel the truth of your new beliefs and comfortable being your new self.

This is how you change beliefs using the masculine approach: you see them for what they are and want them to be. Changing beliefs using this technique takes effort because the unconscious mind needs to see proof it is safe to let go of something that has always been seen as beneficial to your survival. You can do this by constantly and consistently presenting your unconscious mind with the truth and by putting some effort behind it. By acting as though your new way of reacting is who you are now, your unconscious mind will get the message, and eventually, the desired action will feel natural and come naturally.

Let's get into this a little more deeply. Some people think that they cannot change their beliefs because they cannot get themselves to feel the desired belief is actually true. Even though intellectually and logically they may see an old belief as false, it still feels true, and so they think they are powerless to change. This is because beliefs lead to points of view that you possess, and it takes conscious effort to change a point of view. This is likely why you cannot feel what you would like to feel about the new belief you want to adopt. The old points of view are still there at the unconscious level, hanging on, preventing you from releasing the old belief. To let one go, you must admit to yourself, and perhaps even to others you have had conversations with about whatever it is, that you have been wrong—you have been guilty of an error in judgment. In other words, you must identify the points of view and opinions a belief has created and adopt new points of view based on the new belief—thereby replacing the old.

I am living proof that if you really want to, you can change your points of view, and therefore your beliefs through wisdom, knowledge, and your fervent desire to live the truth. As a young man, I was an agnostic bordering on atheist and a confirmed Scientific Materialist. I thought that when you died, that was it, so why not seek pleasure as your number one goal in life. Not so now by any stretch of the imagination. Have you ever seen Animal House? That was me back then.

If this approach doesn't work for you at first, it is because some part of you still isn't completely sure that balance and total empowerment is really what you want. In order to change, you must lose all interest in your old beliefs and the old stories that created them. Think of yourself as the captain of a ship, and you have turned the ship in the direction you want to go. The wake is your past—it's behind you. Leave it there—forget about it. It's gone and will dissipate and dissolve into the ocean.

The truth is, you will never break free from the grip of a belief and point of view until you let it go and forgive whoever and whatever brought it about. That's right—forgive and forget. Uncle Charlie molested you and that's why you fear sex and despise men? Uncle Charlie was a pathetic, dirty old man nobody loved, who is worthy of pity. Forgive him.

Mom said you were lazy and good for nothing? She was angry, she was wrong, and her father was a loser who drank himself into the grave. Forgive her. Holding on to bitterness isn't hurting Uncle Charlie or your mom. It is only holding you back and making you miserable. As soon as you desire empowerment more than anything—even more than revenge or sympathy—as soon as you truly make the effort, cultivate the desire, and jettison old baggage, you will be on your way. Then nothing can stop you. Claim it whether or not it feels true right now. Claim it because you want it to be your reality, and you want to be totally free.

When something in you starts to resonate with your new belief, you will have begun the process of implanting it in your unconscious mind. You must send a feeling-based message to that part of you. By claiming it because you truly want it, that is what you will be able to do.

Chapter Six: Clean Out Your Attic

Give this some thought. As mentioned earlier, you have been around and evolving for a long time—all the way back to the first single-cell creatures that formed in the primordial sea. You are the product of evolution that took place over a mind-boggling 3.77 billion years. Other hominids that evolved along the way branched off onto dead-end paths or developed into chimpanzees or gorillas and such. Some, like the Neanderthals, made it pretty far along the evolutionary path, but eventually could no longer hack it and became extinct when your ancestors, Homo sapiens, came along and took over their territory. But you kept going and going like the Energizer Bunny. You continued evolving until you eventually arrived at the very top of the food chain. There can be no doubt about it at all. You are a member of a very exclusive club—one among the most gifted and highly intelligent creatures that has ever lived. You are the pinnacle of life on earth, and you are now ready to make the leap into the kingdom of heaven.

You can do this because your mind is fantastic. Scientists say we humans typically use only a portion of its capacity. It is the most important tool you have—an amazing tool that is very much like a garden. You can cultivate it, pull out the weeds, water it, plant the right seeds, and allow them to grow. Or you can neglect it, and let it run wild. Either way, cultivated or not, it must and it will bring forth whatever is allowed to grow in it. There may be ideas and beliefs that were planted in your mind while you were going up that are doing you no good. They are weeds that need to be pulled out and cast aside. Plant good, helpful ideas and beliefs in their place. If good seeds are not planted in your mind and allowed to flourish, destructive weeds will take over and will continue producing more of their kind. Just as a gardener or farmer cultivates his plot of land, weeding it, and growing flowers and the fruits and the vegetables he or she wants on the dinner table, so may a person tend his or her mind, weeding out the useless and destructive thoughts and cultivating only those that have the promise of bearing delicious fruit. If such cultivation does not take place, if discipline is not exercised, the result will not be good because what's in your mind will eventually be what's in your world.

As was discussed, over time the outer conditions of a person's life always come to be in tune with his or her inner state. By the process of planting and cultivating positive, constructive thoughts, you will sooner or later discover that you are the master gardener of your mind—the director of your life; captain of your ship of fate. You will also come to understand that your thoughts and beliefs shape your character, which also creates your circumstances. Ultimately, what's in your mind is your destiny.

Something else to know along this line is that the unconscious mind does not understand, or perhaps simply doesn't hear the words "no" and "not." Suppose, for example, you're a tennis player. You're in a big match, it's close, and you are now in a tiebreaker. The next point could decide the match. The point has come for your opponent to serve. As you pass by him at the net changing ends, you say, "You're playing great today, Henry. Don't blow it. This is a big point coming up. Whatever you do, do not double fault."

You've started Henry worrying, and on top of that, his unconscious mind doesn't hear or understand the word "not." All it hears is "do double fault," and it takes that as a directive. Try as he might to do otherwise, Henry will double fault.

Actually, I advise you not to play such a dirty trick. As the saying goes, "What goes around, comes around," and you don't want that sort of negative behavior coming back at you. More will be written about this. The important thing to remember is that self-talk and coaching should always be framed in a positive way.

Think Positive and Rid Yourself of Fear

Let's dig into the issue of fear because fears are beliefs and feelings, and beliefs and feelings are what create your reality.

To learn what you fear, as suggested in the previous chapter, tune into your moment-to-moment stream of consciousness and observe what makes you worried, anxious, resentful, uptight, afraid, angry, and so on. Step outside yourself and identify unsettled emotions, tugs and urges that have become part of your programming. Slow down and consider what triggered a negative emotion. Did your temper flare? Why? Why was it so important for things to go a certain way? If you trace what you felt back to its cause, you might come to a particular variety of fear, and it's been said that only two fears are instinctive: the fear of falling and loud noises. Other fears were acquired, and whatever was acquired can be disposed of.

According to some experts, the fears that hold people back can be grouped under one of six headings:

1. the fear of poverty (or failure),
2. the fear of criticism,
3. the fear of ill health,
4. the fear of the loss of love,
5. the fear of old age,
6. and the fear of death.

I've listed the fear of poverty (failure) first because in many ways it can be the most debilitating. It is self-fulfilling in that traits develop that bring it about. For example, are you a procrastinator? An underlying fear of failure is probably the root cause and can be counted upon to produce the result you fear.

Are you overly cautious? Do you see the negative side of every circumstance or stall for the "right time" before taking action? Do you worry (that things will not work out), have doubts (generally expressed by excuses or apologies about why one probably won't be able to perform), suffer from indecision (which leads to someone else, or circumstances, making the decision for you)?

Are you indifferent? This generally manifests as laziness or a lack of initiative, enthusiasm or self-control.

Step back and listen for internal voices that say "can't" or "don't" or "won't" or "too risky" or "why bother?"

How do you get rid of them? Shoo them away.

Whether you are the president of a company, or a bum

on Skid Row, the only thing over which you have absolute control is your thoughts.

You may say, I can't control what thoughts pop into my head. True. You may not control what thoughts arise, but you can decide whether to discard one or to keep it. You can decide that it is counterproductive and throw it away, or you can turn it over and over in your mind, nurture it and let it grow. Whatever thoughts you keep will expand and eventually manifest themselves.

Beginning now, each time you catch yourself with a negative thought, a thought that says "you can't," "it's not possible," "maybe someone else but not me," get rid of it. Shoo it away.

But you say, "I'm poor, I'm not a good student, I'm not a good salesperson, I'm in the lower third of productivity."

That's your ego talking. You are what you are because of your unconscious beliefs. You want the best for yourself, but your unconscious ego-mind is holding you back because of the way it was programmed.

If what I've been writing about on this page is a serious problem for you, follow the advice given in the last chapter, and go out and buy some self-help tapes that will plant positive thoughts in your mind in place of the negative ones. Play them to and from work and before you go to sleep at night. Use self-hypnosis tapes. Play them over and over for at least a month. Get all that junk out of your head, and replace it with thoughts that are positive.

What about the other fears? They're to be discarded in the same manner. If you suffer from fear of criticism, for example, it probably came about as a result of a parent or sibling who constantly tore you down to build himself up. You'll know this is a problem if you are overly worried about what others might think, if you lack poise, are self-consciousness or extravagant. (Why extravagant? Because of the voice which says you need to keep up with the Joneses.) You must rid yourself of inner voices that tell you to think even twice about what others will say. Take advantage of that fraction of a second between stimulus and response, stop to remember this, and let it sink in. That's how you can change©how to eliminate destructive fears and beliefs.

Let's think for a minute about the fear of criticism. There have been places and times in history when what others thought was worth worrying about. My great, great, great, great, great, great, great grandmother, Susannah Martin, for example, was accused of being a witch, falsely convicted, and hanged in Salem, Massachusetts, in 1692. She was an old lady. Probably, she looked like a witch. But her downfall was the stir she caused after her husband died. She was able to run the farm successfully without a man around. Think of the talk. Such a thing wasn't possible, or so they believed, without the use of witchcraft.

The opinions of Susannah Martin's neighbors mattered a great deal. They led to an unpleasant and untimely death.

What about today?

In Iran, China, Russia, or North Korea one might have to watch out what neighbors think or what the "virtue police" hear, but this simply is no longer a valid concern in developed, democratic countries. What others think or don't think of you or anyone else is their problem. Yet worrying about what they think can cause a great deal of misery, create karma that will have to be worked out, and cause interference between your conscious and your subconscious minds that blocks the channel of communication.

What about the fear of ill health?

To rid yourself of this, it should be enough to know that what you worry and think about is eventually what happens. Ever noticed that it's the people who talk about illness, worry about illness, are preoccupied with this or that possible illness, think they feel a pain here or there or were exposed to some germ, who are precisely the people who stay sick most of the time? The power of suggestion is at work.

How about the fear of the loss of love? This one manifests in the form of jealousy and is self-fulfilling like the others. The person you try so hard to hang onto feels smothered, with the result that you end up pushing that person away. Try being yourself. Give them love, but give them room. It they leave you, they would have done so anyway. You can now move on to a truly meaningful relationship.

Next is the fear of old age. This is closely connected to the fear of ill health and the fear of poverty because these are the conditions a person really is concerned about deep down. The power of suggestion is hard at work here, too. If you think you're too old to do this or that, you will indeed be too old.

Consider this. My children are the same flesh and blood as my wife and me. I saw them being born, still connected by umbilical cords. I clipped the cord of one of them myself. My wife was thirty-six at the time our youngest was born. I was fifty-four. Yet the cells in my body, and in my wife's body, and in my son's body all were the most recent in an unbroken chain of cell division that goes back to the first life on earth. All the cells—my wife's, my son's, and mine—are at the end of a chain that is precisely the same age: billions of years. Those telomeres get shorter and shorter because you think that you should look and feel older as the years go by. The physical body is the overcoat of the mental body. It gets old and decrepit because a person expects it to. It gets older and decrepit because a person stops learning, growing, and playing a role in the evolution of humankind.

When you've learned all you can from this life, the time will come for you to check out. And check out is what you will. No one says you have to be old.

Now we've come to that final bugaboo, the fear of death. As you now have seen, there's nothing to fear except having been fearful in this life. Consider the millions who have had near-death experiences and are no longer afraid to die. They're convinced they'll be greeted by their guides as well as by loved ones who have gone before. They look forward to being bathed once again in the all-encompassing light, which many have described as total, unconditional love. Most do not expect to experience pain. It has been reported by many that the spirit exits the body the instant it looks as though death is inevitable.

Only a handful who have had hellish experiences worry about what they may encounter in the nonphysical world. These folks need to know what you know. Each of us creates his own reality. We experience what we expect to experience, what we think we deserve. In the physical world, this takes time. In the nonphysical world of spirit, which is the medium of the mind, we instantly create our reality, just as we do in dreams. If we expect Hell, the Hell we believe we deserve is the Hell we will get. If we expect Heaven, our vision of Heaven is what we will have.

Anyone who has ever had a lucid dream will understand what I mean. Such a dream is one in which a person realizes he's dreaming. I've had many and I look forward to them because it's more fun than Disney World. As soon as you're aware you're dreaming, you can begin to compose the dream, determine the players, the surroundings, the action. Want to fly over the Grand Canyon? All you have to do is "think" this. Fly over is what you will do, no airplane required. Like anything it takes practice, but I've gotten so I can swoop and turn and loop the loop.

Want to attend a cocktail party populated by Hollywood stars? You'll be there with Robert DeNiro or Julia Roberts. These characters will, of course, be your own thought projections.

You are a dreamer in Infinite Mind's big dream of life, and you can make your waking dream lucid as well. Until now, you may have thought you were at the mercy of conditions outside yourself, that you've either been lucky or unlucky, that chance has brought you where you are. This isn't true. You've brought yourself to this spot, either consciously or unconsciously. If this is not where you want to be, you've arrived because your unconscious mind has been programmed incorrectly, and you are totally out of touch with your Higher Self. Perhaps you hear snippets from it every now and then but ignore what it's trying to say because of other voices which beat it back with "can't," "don't," "shouldn't," "too risky." These are the words of your ego. Your Higher Self wants you to evolve and to enter the kingdom.

Until you started reading this book, you may have thought you were at the mercy of conditions outside yourself, that you have either been lucky or unlucky, and that chance has brought you where you are today. Not so. You brought yourself to this place, and you did so unconsciously. If this is not where you want to be, you will have to change your programming, which likely took place when you were a child.

The Power of Positive Thinking

I suspect you have heard about "The Power of Positive Thinking," that positive thoughts are much more likely to produce good results than negative thoughts. I'm reminded of The Little Engine That Could, an American fairy tale published numerous times in illustrated children's books and movies since its original debut in 1930. The Little Engine was a railroad locomotive that was tasked with pulling a long, heavy train—one that seemed much too large for it—up and over a mountain. But even so, the Little Engine was determined and kept telling itself over and over, "I think I can, I think I can." It was a struggle, but the Little Engine persevered and finally succeeded.

It's a good story and a valuable one to teach young children the benefits of optimism and hard work. The problem is, many of us today were not taught that lesson as children, and in fact, feelings of frustration, discontent, and dissatisfaction were ways of solving problems that many of us "learned" as infants. For example, if a baby is hungry, he or she expresses discontent by crying. Lo and behold, a warm and tender hand appears magically out of nowhere and brings a bottle of milk. Later on, if the baby is uncomfortable, again, he or she will again express dissatisfaction, and the same warm, comforting hands magically appear and solve the problem. That's fine for babies but, unfortunately, many children continue to get their way and have their problems solved by indulgent parents merely by continuing to express their feelings of frustration when things don't work the way they want. All they have to do is feel frustrated and dissatisfied, express their dissatisfaction, and the problem will be solved. Sometimes what have become known as "helicopter parents" continue to cater to their children in this way all the way through high school, college, and beyond.

This way of life "works" for infants, and for some children. But it does not work in adult life when a person is out in the world on his or her own. Yet many continue to expect, perhaps unconsciously, that it will work. They seem to think that by feeling discontented and expressing their grievances—if only they feel put upon enough—life, or someone will take pity on them, rush to their aid and solve their problems. Let me assure you that 99.99 percent of the time that is not going to happen. It is my advice that you take responsibility for every aspect of your life.

Imagine, for example, you land an entry-level job as a management trainee in a big corporation. With you in training are several other bright young men and women fresh out of business school. Imagine the way things work in this company is often not to your liking. Management trainees, for example, are relegated to cubicles with five-foot-high walls affording little or no privacy, while the senior staff all have corner offices with large windows and spectacular views of the East River. You spend a good deal of time grumbling to yourself and to others about this injustice, subconsciously believing that will get you out of that cubicle and into a corner suite. Your fellow trainees, on the other hand, spend their time making positive suggestions and anticipating and providing for the needs of customers as well as for fellow workers higher up on the corporate ladder. Whom do you suppose is most likely to be first to break out of his or her cubicle? The one who constantly complained? Or the one that consistently delivered the goods?

Don't you feel a twinge inside that intuitively "knows" the positive attitude, the attitude of service to others, will inevitably win the day? That "twinge" is a message from something inside you that knows the correct answer called "intuition."

If you have been ignoring that feeling when it comes, now is the time for you to begin recognizing such messages. They have a light and airy feeling to them, even though they may seem to run counter to egocentric notions, such as, "The first order of business is to look out for number one." That egocentric notion may work in the short term, but in the long term, it is bad advice. The fact is that it's always best to under-promise and over-deliver to customers and bosses—as well as to anyone else for that matter. By over-delivering, your reputation grows as you create positive vibes and positive opinions of you by those with whom you come in contact. A reputation that you are someone who can be counted on can only lead to good outcomes and opportunities for you in the long run.

Let's consider for a moment why some people may spend their valuable time on earth grumbling and complaining away opportunities to get ahead. It's often because they have felt frustrated and defeated for so long—ever since they were babies in a crib and while growing up with indulgent parents—that those feelings have become ingrained. Their minds are in a kind of holding pattern, and it's never occurred to them to step outside of themselves in order to get in touch with intuition that would tell them, if they would only listen, that grumbling and complaining are counterproductive and accomplish nothing. Until they wise up, they will continue—to their own detriment—to radiate those feelings, and as sure as night follows day, their discontent will lead to failure.

No matter what your mindset, if you want to change it, it's important to realize, as has been discussed, that beliefs and feelings are intertwined. It might be said that feelings are the soil in which thoughts and ideas grow. If you are habitually grumpy and in a bad mood, you need to lighten up and begin seeing the glass half full. Moreover, when you begin working toward a goal, try thinking how you will feel when you reach it—and then actually make yourself feel that way. I'm serious. Conjure up the feeling of "Success!" The thrill of accomplishment will communicate the belief to your unconscious mind that it's inevitable you are going to achieve what you have set out to accomplish. The feeling creates the belief, and the belief creates the feeling. A mental model of success will be etched into your unconscious, and that the desired outcome will surely come about.

Let's say you are pursuing a challenge and fervently want to accomplish it. Assuming you have the education, the knowledge, and the qualifications required to reach your goal, and assuming you feel strongly about it at an emotional level, you will almost certainly realize success. This will happen because your unconscious mind, which is connected to the universal subconscious and all other minds, will go to work and act like a magnet, drawing to you what you need. The greater your desire, the more your unconscious mind strives to produce results.

In summary, belief and emotions are the keys. It's important to feel the joy of having accomplished what you set out to accomplish before it actually happens. This will convince your subconscious the goal has already been reached and the universal mind will cause it to be reached as a result.

Be Likable and Appealing to Others

Leaving psychological suffering behind will require effort and work, and that means things will be easier if you have help along the way. The most likely source of that help will be friends, partners, and mentors. Obviously, the going will be easier and you will attract more help if people like you and want to work with you. Therefore, it should go without saying that it's important to be someone others want to be around—someone people would like and want as a friend. That means you need to be someone who "talks the talk," and "walks the walk." Perhaps you know a person who does the opposite. If not, you are likely to come across someone like that in your business dealings, so be prepared and never, ever be one of them. In public such people talk openly—some even brag and boast—about the importance of having integrity and doing the right thing. But in private it's a different story. They bad mouth people and do things that aren't consistent with the honest-John public persona they hope to project. People quickly see through these phonies. As the old saying goes, "Say what you mean, and mean what you say," and people will respect you for doing so.

Obey the Laws of Physics

If you are a Materialist and think matter is all that exists—that there is no God and nothing spiritual—you might come to the conclusion you can do whatever you want, harm whomever you want, and never have to suffer any consequences. But that is not how things work. I agree it may work for a while, but eventually you'll get back what you gave out because, just has there are laws of physics, there are laws of metaphysics. People must obey them, and nations must as well. Take Japan and Germany in World War II as examples. Both countries had astonishing victories in the beginning. Each country benefited from a national zeitgeist that they were invincible. But they ended up being crushed because the atrocities they committed came back upon them with a vengeance..

Always Keep Your Life in Balance

You are probably familiar with the ancient Chinese symbol composed of a white "yin" interlocking a black "yang" that represents the dual nature of things. It symbolizes that we live in a world that is composed of opposites: Up, down, black, white, good and evil. Without the tension opposites create, nothing would or could exist—everything would fall apart. Follow the advice of this book and enter the kingdom but do not allow complacency to set in. Always seek new challenges, realizing that without one, self-destruction may result. Always strive to continue growing.

It can also be comforting to know there can be no growth without at least some discontent. Deep within, you know what is best for you. There is an urge built into you that pushes you to strive for growth, and for most of us, growth will not continue without some agitation and discontent. So study your dissatisfactions. They will tell you what you are about to leave behind and possibly point you in a new direction. Be willing to be uncomfortable. It is the way to grow.

As you contemplate your future course, it is also important to realize you can only attract that which you feel worthy of. Self-esteem is critical to success. That's why I urge you to get rid of the psychological baggage. The truth is you are not what you have, and you are not what you do. Beneath your fear and negative programming, you are perfect—an enlightened soul, fully self-actualized and a living example of unconditional love. The more you can let go of fears, the higher your self esteem will be, and the more options you will have and more risks you can take. The more you like yourself, the more others will like you, and the more worthy you will feel.

You can have anything you want if you can give up the belief that you cannot have it—assuming what you want does not conflict with someone else's belief. If, for example, you desire a fulfilling, one-to-one relationship, but demand it to be with a particular person, you are not operating in harmony with the universe.

Another example is in the area of accomplishment. You must get the education necessary to create what you want. "Where your attention goes, your energy flows." You attract what you believe you are and that which you concentrate upon. If you are negative, you draw in and experience negativity. If you are loving, you draw in and experience love. You can attract to you only those qualities you possess. So, if you want peace and harmony in your life, you must become peaceful and harmonious.

Something else to understand is that a stronger emotion will always dominate a weaker one. Every idea can be the beginning of a manifestation—although unless you nurture it, think about and develop feelings about the idea, it will not become expressed in reality. It does not matter which idea you consciously favor, even know to be desirable, a stronger emotion will nullify a weaker one, and the strongest emotion will begin to permeate all aspects of your activities. For example, if you are emotionally focused on the sexual desirability of a particular person, you may begin to create circumstances that will increase the likelihood of an eventual sexual union.

It is also important to realize that new information you accept into your mind will destroy previous information of a similar nature. Once a pathway of information has been created in you, a new viewpoint will develop and prevail unless new information comes in to replace and destroy it. Let's say, for example, you fall off and get hurt while horseback riding. That may be the end of your experience with horses because you will have just been programmed negatively about horseback riding. This is why instructors always urge new riders to climb back aboard immediately. You need new, positive input to erase the trauma of the fall.

The mind is engaged in an endless state of growth and reorganization. As a result, it is possible to reprogram yourself. You can do this by using the feminine or masculine techniques described previously, or some combination of the two, and by reinforcing new beliefs and points of view by repeatedly listening to success-meditation recordings and using visualization techniques. If you feel anxiety in crowds, imagine yourself relaxed in a crowd of people. When you fear doing something, and yet have the courage to do it anyway, you will soon do a mental flip-flop and may even become addicted to doing it.

Here is a case in point. Suppose you fear skydiving, or skiing fast almost straight down a steep mountain. If you force yourself to do so anyway, the experience will release endorphins, which are produced by the central nervous system and the pituitary gland and can produce a feeling of euphoria very similar to that produced by opiates. The result can be that you become somewhat addicted to skiing fast down mountains and skydiving.

You have within you everything required to make your earthly incarnation a paradise if you choose to accept that which is your divine birthright. We live in a universe of abundance, although the majority of humans populating our planet appear to view it as a universe of scarcity.

Heed the call. Take the leap. But do not go off half-cocked. Plan it out. Take a full day. Take more than one. Take as many as necessary to develop your plan.

Thank you for taking this journey with me. I hope you enjoyed it and feel you got something important from it. If so, do yourself, others and me a favor. Pass this book along to someone you know, someone who might benefit from knowing the truth. And tell others about it. As more and more people learn the truth—that we are all one life and that hurting others only hurts ourselves—life on earth has got to improve. Wouldn't it be great if the bickering and end fighting were to cease, or at least to recede? It can happen, and it will happen, if enough people read this book.

# # #

About Stephen Hawley Martin

Stephen Hawley Martin is a former marketing executive and consultant and the author of more than three-dozen books, including five published novels, half a dozen business management titles, and quite a few self-help books and metaphysical investigations. He is a former principal of the world-renowned advertising agency, The Martin Agency, the firm that created the GEICO Gecko and "Virginia is for lovers." Today, Stephen is editor and publisher of The Oaklea Press. Listed in *Who's Who in America,* and best known as a award-winning author, Steve is the only three-time winner of the *Writer's Digest* Book Award, having won twice for fiction and once for nonfiction. He has also won First Prize for Visionary Fiction from *Independent Publisher* and First Prize for Nonfiction from *USA Book News.* He and his wife of 35 years live in central Virginia.

To get in touch with Stephen and learn about other books he has written, visit his website:

www.shmartin.com

This two-book volume contains the bestselling title, *Life After Death, Powerful Evidence You Will Never Die* and the sequel, *Heaven, Hell & You.* As one reviewer, a medical doctor, wrote: "Extraordinary findings . . . will keep readers on the edge of their seats as they burn through this well written book's pages."

Kindle: ASIN: B07J46QQW8
Paperback: ISBN-10: 1727782038

Some searching for the Grail thought it was the cup Jesus used at the Last Supper. Others, a dish, or a stone. All believed it had powers to create happiness and wealth. This book explains that the True Holy Grail is secret knowledge that enabled Jesus to work miracles.This book tells that secret.

Kindle: ASIN: B07MYQ8N8V
Paperback: ISBN-10: 1794500715

You may not know it, but you have a powerful cosmic mind at your core, and you can learn to access it. From time to time, you likely have been in touch with this part of you, although perhaps only fleetingly. There's a reason it rarely happens. Your ego is getting in the way. This books explains how to push your ego aside and get in touch.

Search: ASIN: B09J7MBBL3

All the knowledge of the universe resides within you because at a deep level all minds, past and present, are connected. Everything that has ever happened, every thought, every idea is there. The trick is to draw out information when you need it. In this book Stephen explains how he learned to do so and how you can, too.

Kindle: ASIN: B07HHFFWP8
Paperback: ISBN-10: 1723835250

This 5-Star rated, fast-paced thriller is based on the true nature of reality as revealed in the book you hold in your hands. The heroine travels to the Caribbean Island of Martinque to save her father and learns the secret of life in the process. A page-turner, this novel won First Place for Fiction from *Writer's Digest* and First Place for Visionary Fiction from *Independent Publisher* magazine.

Kindle: ASIN: B08S7MG4WM
Paperback: ASIN: B08SB6VG9L

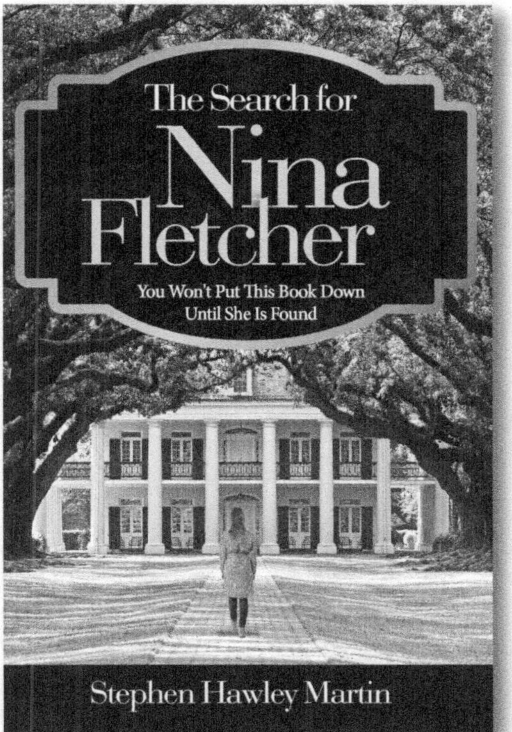

In this romantic suspense thriller, Rebecca wants to save the beautiful plantation home where she grew up, but to do so she must find her mother. If only she could remember what happened in the basement of the old house in Baltimore long ago. She must find out what happened there, she must!

Kindle: ASIN: B01J6MQZXS
Paperback: ISBN-10: 1535580879

Made in the USA
Monee, IL
13 September 2026

62975577R00059